THERMODYNAMICS:
FUNDAMENTALS, APPLICATIONS

THERMODYNAMICS:

Fundamentals, Applications

OTTO REDLICH

Principal Investigator, Lawrence Berkeley Laboratory,
and Lecturer, Department of Chemical Engineering,
University of California, Berkeley, Calif., U.S.A.
former Professor, Technische Hochschule, Vienna, Austria

ELSEVIER SCIENTIFIC PUBLISHING COMPANY
AMSTERDAM - OXFORD - NEW YORK 1976

ELSEVIER SCIENTIFIC PUBLISHING COMPANY
335 Jan van Galenstraat
P.O. Box 211, Amsterdam, The Netherlands

Distributors for the United States and Canada:

ELSEVIER/NORTH-HOLLAND INC.
52, Vanderbilt Avenue
New York, N.Y. 10017

ISBN: 0-444-41487-8

Printed in The Netherlands

You shall distinguish between Truth's well-rounded, unshakable core and current opinions, that you cannot rely upon.
(Free translation from Parmenides, about 500 B.C.)

PREFACE

Thermodynamics, like Helen of Troy, has always been "admired much and much reviled" (Goethe, Faust, Second Part, Act 3). There has been ample reason for both admiration and revilement.

In the ideal image, thermodynamics is the paragon of clarity and infallible strictness. Reality has not entirely lived up to this image. Sometimes one gets the impression that thermodynamics is an obscure collection of special tricks. Even such simple concepts as extensive and intensive quantities have been misunderstood and distorted quite often. For some time, highly elaborate attempts have been made to introduce mathematical methods into thermodynamics. They have not made any contribution to eliminating superficial or deeply rooted errors.

The most serious deficiency in traditional thermodynamics has been the lack of definition of the basic concepts. It has long been recognized that work is always obtained by integration of one work factor ("generalized force") with respect to another one ("generalized coordinate"). The names apparently have been borrowed by Zemansky from analytical mechanics, but lists of the work factors had been presented by Helm already a hundred years ago. Until recently, however, only a single scientist (Ehrenfest, 1911) has ever asked what the characteristics of these two classes of properties are, and his search was by no means successful. Such fundamental concepts of thermodynamics are supposed to be applicable to all other sciences; they must be well-defined and clear beyond any doubt.

The lack of definition of generalized coordinates and forces entails immediately a similar vagueness in the concepts of work and energy. G. N. Lewis, with his infallible flair, felt that something was wrong. He commented (1923): " ... and indeed an unfriendly critic might claim, with some reason, that the law of conservation of energy is true because we make it true, by assuming the existence of forms of energy for which there is no other justification than the desire to retain energy as a conservative quantity." Neither this remark itself nor the subsequent emphasis on the practical justification of the energy law actually settles the issue; but a serious fundamental problem takes shape in this seemingly flippant comment.

The method of constructing these and a few other basic concepts has been developed in the last fifteen years. It starts by searching for the characteristics of those concepts that we inevitably need in the quantitative description of our observations.

This procedure, which may be called the "epistemological method", resolves a puzzling question: Why do the generalized coordinates and forces occupy exceptional positions among the innumerable quantities introduced in physical sciences? It explains why, for instance, the electric charge is a generalized coordinate, the voltage a force, the electric conductivity neither. There is no room left for any arbitrariness. By establishing explicit, clear and inevitable criteria for coordinates and forces, we remove any vagueness from the concepts of work and energy. This development is coherently described in the present book.

If we wish to describe the whole path from the fundamentals to the practical applications, we have to show the use of thermodynamic principles in various sciences and in technology. Chemical applications are described here in some detail. The methods have been designed by Lewis and Randall, and further developed by Pitzer and Brewer and other authors. Practical rules are demonstrated in examples, some of which are the fruit of work at Shell Development Company.

The author desires to lead the reader to a clear understanding without gaps. This intention is believed to justify the inclusion of some primitive discussions of a few rather remotely related topics in the Appendices. A few general comments on computation methods may be useful. The expansion of automatic computation invites comparison of the efficiency of tabular, graphic and algebraic methods. Computational automation should not lead to the belief that the critical judgment of the observer or compiler can be replaced by the "objective" method of Least Squares. Actually this method only shifts the personal judgment to the choice of statistical weights, an important point in thermodynamic computations.

This book is designed to present the essential tools for independent application of thermodynamics not only to the chemist and chemical engineer, but also to the physicist, to the electrical and mechanical engineer, and to the biochemist. An effort has been made to achieve brevity as the result of precise language, not by skimping on derivations.

An adequate notation has always been a problem in thermodynamics. Essentially, the notation of Lewis and Randall is used, but all molal quantities (not only partial molal ones) carry a bar. Changes in the meaning of a symbol are sometimes inevitable. They should caution us against too rigid a link between concept and symbol.

The author profoundly appreciates the advice and help extended to him by Professor Leo Brewer. His sincerest thanks are due to Dr. Constantine Tsonopoulos, who carefully revised the manuscript and contributed numerous and important critical comments. He gratefully acknowledges the frequent advice of Dr. Fred H. Stross. He appreciates the untiring care with which Miss Karen M. Krushwitz prepared the manuscript.

The author gladly takes the opportunity to express his gratitude to the Lawrence Berkeley Laboratory and to the Department of Chemical Engineering, University of California, Berkeley, for the use of their research facilities.

Berkeley, California, 1 May 1976 Otto Redlich

CONTENTS

CHAPTER 1. THE BASIS OF THERMODYNAMICS

A new cultural epoch, as a rule, finds an early expression in architecture. The ideas developing in the arts, sciences, philosophy, and politics usually are manifest later. The culture of our time is developing in the direction indicated by Frank Lloyd Wright, Gropius, Loos, Mies van der Rohe. In their buildings, the plan is simple and clean. We recognize immediately the load and the load-bearing members. The shape of each part expresses its function. There are no dirt-catching niches, corners and cornices.

Such a model we need in constructing the building of thermodynamics. Understanding thermodynamics means that we are able to retrace any statement, any application step by step to more general ideas until we arrive at the basic principles. For this reason we need a clear and unambiguous construction.

It is the ability of retracing to the fundamentals that distinguishes the technologist from the technician, who applies well established formulae according to old examples. Whenever a really new problem arises and the established methods fail, the technologist has to go back step by step in order to find the point of departure from where he must start to blaze a new trail.

1.1 THERMODYNAMICS AND THE PHYSICAL SCIENCES

For historical reasons, traditional thermodynamics has not been very close to our ideal building. Carnot started from a special problem, the maximum efficiency of a steam engine. He solved it brilliantly for a much wider field, namely, for any heat engine. Similar expansions of scope went on and on. Various kinds of mechanical, electric, magnetic questions, many chemical and also biological problems have been elucidated. The expansion process has not stopped up to now.

But the numerous additions to the building of thermodynamics, the frequent repairs and extensions did not promote the clarity of construction.

In recent times, the opinion has been widespread that improvements in the mathematical methods will produce the desirable clarification. Actually the building does not need a strengthening of the mathematical pillars and cross ties but the base, the underpinnings, must be replaced. Then the construction of a clean and functional building presents no serious difficulty.

1.11 Equilibrium. There is a good reason for examining the basis of thermodynamics very carefully. It is the role of thermodynamics in the family of physical sciences. We apply the methods of thermodynamics in full confidence in the study of muscle action as well as surface tension, in the theory of the development of stars as well as in the description of magnetic phenomena. To be sure, we often transfer results of other branches of science. For instance, we realize that the propagation of neural impulses is an electric phenomenon and make use, of course, of our knowledge of electricity. But it is a common feature of all branches of physical science that is elucidated in thermodynamics. Which are the concepts and terms that are common to all branches and therefore should not be derived from any single one?

The common feature is the state of equilibrium. The world as a whole and many of its parts are unquestionably very far from equilibrium. Yet there are many aspects in which parts of nature can properly be described as being in equilibrium. Thus the idea of equilibrium appears in many branches of science. The state of equilibrium therefore must be described in an entirely general way. This is the topic of classical thermodynamics. The base of thermodynamics is provided by the concepts describing equilibrium and its attainment, concepts common to all physical sciences.

These sciences branch out from the common root in problems of kinetics, dynamics, molecular theory and so on. Thermodynamics itself sprouts beyond the classical problem in statistical molecular theory and the description of irreversible processes.

1.12 Science, Pure and Applied. There is the ancient story of a man who described a marvelous jump he had performed a year before on the island of Rhodus. His bluff was called by the shout "Let this place be Rhodus, jump here!"

It is only on the stage that every night the murderer of King Henry VI and the Prince of Wales can woo and win the Prince's mourning widow [11]. The actual historic event cannot be repeated. But a nuclear chain reaction can indeed be repeated in awesome reality anytime. Reproducibility distinguishes the physical sciences from history and other intellectual endeavor. Here we can call the bluff. More important, honest errors are corrected or -- more often -- they silently disappear from the scene [12].

Two consequences follow immediately from the requirement of reproducibility in the physical sciences. The description of an event is complete only if it comprises instructions sufficient for a reproduction of the event. Moreover, we must be able to put these instructions into reality. In order to prepare the event we must know how to change things within a sufficient range according to our pleasure. To make an omelette we must have the recipe, we must also know how to mix milk and eggs, and how to heat the mix without burning it. On the other hand, the most eloquent description of nectar and ambrosia, the food of the Olympian gods, does not enable us to prepare supper.

This requirement shows that there is no essential borderline between pure and applied science. Technology, after all, rests on the art of manufacturing a thing or realizing a process according to our pleasure. But this is also an intrinsic part of science itself because it is needed for reproduction of a described event. When we build a nuclear reactor, it makes little difference whether we shall investigate the structure of the nucleus or use it for electric power: in both cases we need information of preparing the nuclear fuel, of construction, of methods for controlling the reaction and so on. The distinction of pure and applied science is neither essential nor sharp.

1.13 The Definition of a Quantity. The importance of quantitative measurement in our civilization need not be emphasized. But what is the common feature of all our diverse measurements? To answer questions of this kind we shall often examine a primitive example and extract from it the general characteristics.

We choose a practical example: Does the rug we see in the store fit the length of our living room, or is it shorter, or is it longer? We can answer the question by spreading the rug in the room, or indirectly by measuring the length of the room and the rug in the usual way with the aid of a meterstick.

"The usual way" of measuring the length of a rug contains all important features of any measurement.

We establish the concept of length by distinguishing between the three cases of the rug being longer, shorter, or fitting the living room. We arrive at a decision by comparing the rug and the living room. The specific manner of comparing is trivial in this case. Important is the fact that we have to compare two things. Generally we create the concept of a quantity by prescribing the method by which we recognize the three cases (a) $A > B$; (b) $A < B$; (c) $A = B$. As long as we have not such a method, we do not have a concept of length. Conversely, the measuring method is the only essential feature. The following details can always be added without difficulty.

In order to assign a number to the length of the rug, we arbitrarily choose a standard or gauge, in the present case a meterstick. We compare the length of the rug and the meterstick by laying them side by side, repeatedly applying the stick and using fractions of the stick. The result is expressed by a decimal number. The standard meter is compared in several steps with the primary standard, a platinum rod in Paris, which had been arbitrarily chosen.

The assignment of a number is not necessarily restricted to a meterstick. We could also take a yardstick, divide it into three parts, each part in twelve parts, each part again into 4 parts, and so on; but who would use such a cumbersome procedure? Actually we have still more freedom in the assignment of numbers. We could divide our stick in a logarithmic scale, or by a square root scale, or by a scale prescribed by any monotonically increasing function. The restriction of the function is necessary to maintain the "greater" and "smaller" relations in the various scales.

All this applies to the definition and measurement of any and every quantity. We determine the weight by comparison with standard weights, the hue of a color by comparison with standard colors, a pressure by comparison with a gauge.

The result is a rational number, i.e., a decimal number with a finite number of digits, rarely more than 8 digits (length, weight, electrical resistance), never more than 11 digits (time).

Conversely, when we wish to define a quantity we have to set up an experimental method for comparing two things in such a manner that a property of the first thing is found to be greater than, smaller than, or equal to the property in the second thing -- this is sufficient to define the property as a quantity. Only a few arbitrary conventions regarding gauge and scale calibrations must be added.

1.2 BASIC CONCEPTS

1.21 How Do We Construct a Basis? In the history of thermodynamics the question of the basic concepts has never been clearly faced. It is a peculiar problem. We have to forge the tools that are to be used in thermodynamics and in many other branches of science. Obviously we have to develop the basic concepts without taking recourse to any of the branches.

Traditionally this has not been done. Concepts such as work and energy have been taken over from mechanics and used in all other branches without any concern whenever the scope of thermodynamics has been extended to a new field.

If we wish to avoid such a sloppy procedure, the question remains: How do we find concepts so general that they can be used in all parts of science? Moreover, how can we be sure that they will be general and always applicable even to new branches of science?

The answer has been given by Immanuel Kant, the German philosopher, two hundred years ago. From the long and involved discussions of his "Critique of Pure Reason" we can distill out the plain but important result: generally valid are those concepts that are indispensable for the description of the world. A concept that is necessary for expressing any and all of our observations is automatically valid and applicable in all physical sciences [13].

It goes without saying that this principle never leads to a substantive statement about the world. It provides use with "rules of order" for investigating, or, if you wish, with a frame for our picture of the world.

1.22 Object and Isolation. Let us set aside for a moment all we know about the world. Then we look around and ask how we shall describe what we are seeing. It is obvious then that we have to go at it piecemeal. We have to take a part of the world for a description, then another, and so on. In order to complete the description of each individual part, we have to keep it unchanged.

From this simple discussion we extract two important concepts, object and isolation. We call "object" any part of the world that can be kept unchanged, and we call "isolation" the circumstances that ensure that the object is unchanged whatever may happen in its environment. The two concepts are coupled: An object is something that can be isolated, and isolation can be discerned only by observation of an object and its environment. An example is a strongbox: its contents are not changed by fire or flood or an earthquake.

Frequently we are satisfied to talk of isolation if it is established only in regard to some specific influences, as in a calorimeter or a thermos bottle.

1.23 Idealization. At this point we interrupt the flow of discussion in order to consider a practically important matter, namely, idealization. In thermodynamics, as in all sciences, we use idealized concepts. Sometimes we specify the idealization in the name ("ideal" or "perfect" gas), often we do not. Any calorimeter has solid supports, wires leading to the environment, and perhaps other devices crossing the vacuum jacket. In a well designed calorimeter these imperfections have only a small influence on the result; moreover, we try to estimate their magnitude and to correct for them. The uncertainty of this estimate then is part of the experimental error. The accuracy of an observation is limited not only by the reading error but by numerous circumstances, such as deviations from ideal behavior of the instruments used, imperfect purity of the substances, and so on.

The justification of using idealized concepts and results is based on the assumption that the effect of deviations from ideal conditions decrease with the magnitude of these deviations. This assumption is almost always correct but there are exceptions, such as the influence of a catalyst.

1.24 Interaction. After the description of isolated objects, our second general problem is the interaction between two objects. Our interest in this question is not only theoretical but also eminently practical. It is by establishing interaction with a second object that we change an object from its present to a predesigned state.

Every day, from the moment we step out of bed, we observe and participate in hundreds of different interactions. As before, we ask what the common features of these interactions are. For this purpose we shall consider a few simple types of interaction and again extract the general features.

The simplest example is shown in Fig. 1.1: a weight hanging on a spiral

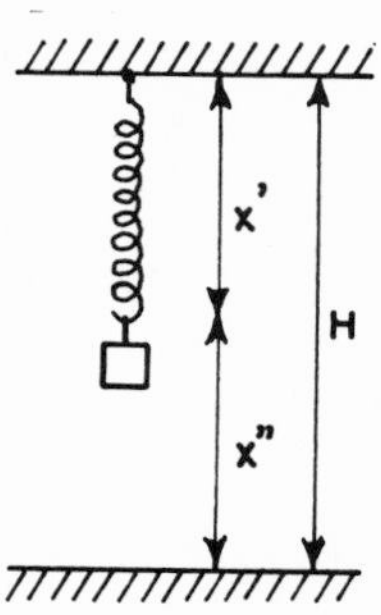

Fig. 1.1. Spring and Weight

spring. For the purpose of interaction the spring is extended to the length x', at the same time the weight must have been lifted to the height x''. Interaction can take place only if the sum of the two variables of the two objects is equal to the height H of the room so that

$$x' + x'' = H \tag{1.1}$$

is a condition of interaction, necessarily satisfied whenever two objects interact.

Similarly, movement of the piston in Fig. 1.2 and expansion and compression of two gases in the volumes V' and V'' can take place as long as the total volume V is not changed, so that the interaction condition is

$$V' + V'' = V. \tag{1.2}$$

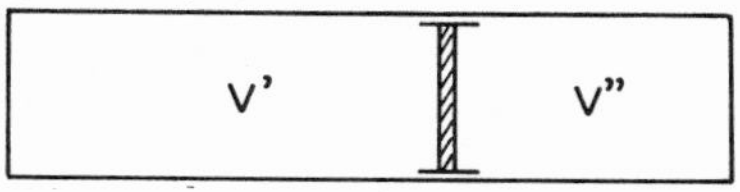

Fig. 1.2. Volume Interaction

If we connect a capacitor with a galvanic cell, some electric charge may flow from one to the other but the total electric charge cannot change. If we introduce water into an evacuated glass bulb some of it will vaporize, but the sum of the amounts (number of moles) in the liquid and in the vapor remains unchanged.

What are the common features of these and all other kinds or modes of interaction?

(1) Interaction is established by some appropriate device such as the hook and eye in Fig. 1.1, the piston or movable membrane in Fig. 1.2, the pair of connecting wires and the switch in the electrical example, a catalyst in starting a chemical reaction and so on. Conversely we can prevent interaction or isolate the two objects from each other with respect to a mode of interaction: in the first example by releasing the hook, in the second by drilling a hole sidewise into the piston and fixing it by means of a pin, in the third by opening the switch, in the fourth with the aid of an impermeable membrane, and so on.

(2) The characteristic feature of any interaction is the loss of one independent variable. Before connecting spring and weight (Fig. 1.1), the variables x' and x'' were independent. We could extend or compress the spring

without interference by the weight, and we could lift or lower the weight without regard to the spring. After interaction is established, however, only one independent variable is left, due to condition (1.1). As soon as we lift the weight, we compress the spring.

(3) We realize now in these examples that we can change an object by interaction with another. We can extend the spring or charge the capacitor, and so on.

The interaction condition need not be formulated as simply as in (1.1) or (1.2). If we had introduced the quantity

$$Q = e^{V} \tag{1.3}$$

in the place of the volume, we could describe all objects in the world just as well as by means of V. But the interaction condition would be

$$\ln Q' + \ln Q'' = \ln Q \tag{1.4}$$

$$Q'Q'' = Q \tag{1.5}$$

with a constant value of Q. In general, we choose our variables so that the simple form (1.1) or (1.2) applies, but this is only our arbitrary conventional choice.

It is important to realize that any number of similar examples could be found in the various fields of physical science. Whichever example we consider, we shall always have to consider <u>interaction</u>, <u>establishing</u> interaction, <u>mode</u> of interaction, and <u>condition</u> of interaction.

1.25 Coordinates. In the choice of variables for the examples of the preceding section we had a considerable leeway. In the second example we might have taken instead of the volume either its exponential function or its square root. But the density of the gases or their refractivity or their viscosity would have done just as well, though the interaction condition would have been more complicated.

But as soon as the state of an object depends on more than one variable, the choice is seriously restricted.

As an example we take the capacitor-balance (Fig. 1.3). The state of the capacitor can be varied in two independent ways: we may lower or lift the upper plate of the condenser by manipulating the balance,and we may withdraw some charge from the capacitor or add some. These interactions are independent and we can inhibit either one according to our pleasure: We may arrest the balance, or we may open a switch in the capacitor leads.

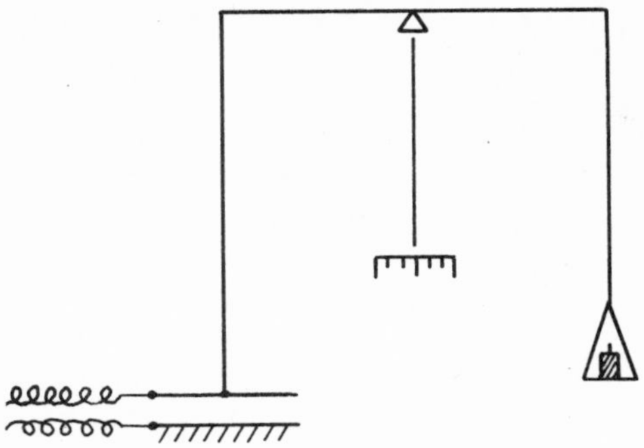

Fig. 1.3. Capacitor-Balance
(This figure and Fig. 1.1 has been taken from Ref. 13)

The mechanism of each of the two modes of interaction endows one variable with a particular significance, namely, the height of the upper plate and the charge of the capacitor. They are the variables that can be arbitrarily kept constant. Each also can be changed arbitrarily without affecting the other. Such a behavior is generally called orthogonal. A set of independent variables that are orthogonal on one another will be called a set of generalized coordinates.

Orthogonality is a very special quality of a variable, determined by the nature of the mode of interaction. The voltage of the capacitor cannot be a coordinate because we cannot ensure its constancy by preventing electrical interaction while manipulating the balance. To be sure, we can maintain a constant voltage but only by interaction in both modes according to a certain condition, not by eliminating one of two modes.

The practical significance of orthogonality is obvious. We need it for the description of nature as well as for changing the state of an object according to our pleasure. There are many more variables (such as the electrical conductivity or an extinction coefficient) that could be used as a member of a set of independent variables. But they could not be kept constant while the object interacts in various modes. For generalized coordinates we must require that any of them can be kept constant while other coordinates are arbitrarily changed. The choice of generalized coordinates is tied to the nature of each mode of interaction involved.

The mechanical connection illustrated by the example of the spring and the weight leads to a spatial coordinate as a generalized coordinate. Another spatial coordinate is the angle of rotation in the case of a rotating object.

The volume is the generalized coordinate prescribed by the use of a piston (or any movable wall) as the interaction mechanism. The electric charge satisfies the conditions for a coordinate as mentioned before. For magnetic interaction the coordinate is the magnetic induction. This list is not exhaustive; it should not be since the door must always be kept open for an entirely new kind of interaction.

1.26 Forces. It is entirely correct to say that we must exert a force on an object (directly or through another object) if we want to change its state. This is also a pertinent remark since the ability of arbitrarily changing the state of an object is essential (Section 1.12).

But a remark of this kind is of course very far from presenting a definition, which after all must teach how to measure a force. We shall arrive at a definition if we return to the discussion of interaction (Section 1.24) and further observe the play and counter-play of two objects. As the first example we choose again the mechanical interaction of a spring and a weight (Fig. 1.1). After hooking the weight into the end of the spring and thus establishing interaction, we observe what happens now. Obviously there are three possible cases:

(a) The weight goes down so that the spring is extended

$$dx' > 0; \qquad dx'' < 0. \tag{1.6}$$

(b) The spring contracts so that the weight is lifted

$$dx' < 0; \qquad dx'' > 0. \tag{1.7}$$

(c) Neither spring nor weight change

$$dx' = 0; \qquad dx'' = 0. \tag{1.8}$$

Now we define the quantity force by saying that the force f'' exerted by the weight is greater than the force f' exerted by the spring in case (a), or that in the three cases

(a) $f' < f''$ (1.9)

(b) $f' > f''$ (1.10)

(c) $f' = f''$. (1.11)

The third case is called equilibrium between the two objects.

This is essentially a complete definition of the mechanical force. We have only to choose a standard force. For this purpose one chooses the standard kilogram (a piece of a platinum alloy kept in a vault in Paris) as a standard of mass and sets up rules for multiples and fractions of this mass. The standard gravitational forces are the products of the masses by the gravitational acceleration (which depends on the location, but at sea level is close to 9.80665 meter$\cdot$sec^{-2}).

The really important conclusion from our discussion is the fact that it can be immediately generalized to any possible interaction. The observation of the interaction of two gases by means of a piston (Fig. 1.2) leads to the definition of pressure. We call pressure (conventionally multiplied by -1) the generalized force conjugated with the volume. The observation of the flow of the electric charges between a galvanic element and a capacitor leads to the relations (1.6) to (1.11) if we understand by x' and x'' the electric charges on the two objects and by f' and f'' the voltages. Thus we define the voltage or electromotive force, being the generalized force conjugated with the charge. The magnetic field strength is the generalized force for the magnetic induction, the surface tension for the surface area, and so on. Our way of defining is applicable to the interaction between deuterons and helium by fusion, though the equilibrium has not yet been realized. It is applicable to any new mode of interaction, which may be found tomorrow.

Generalized forces are measured by comparison of an unknown force f' with a standard force f''. For the quantitative determination, equilibrium between the two objects must be established, at least with respect to the interaction involved. An example is shown in Fig. 1.4. The flowing water is certainly not in a state of equilibrium but there exists equilibrium between the water at the tap points and the liquid in the manometers. In each branch of physical science special methods for the measurement of forces may be developed but they are always based and calibrated by measurements at equilibrium.

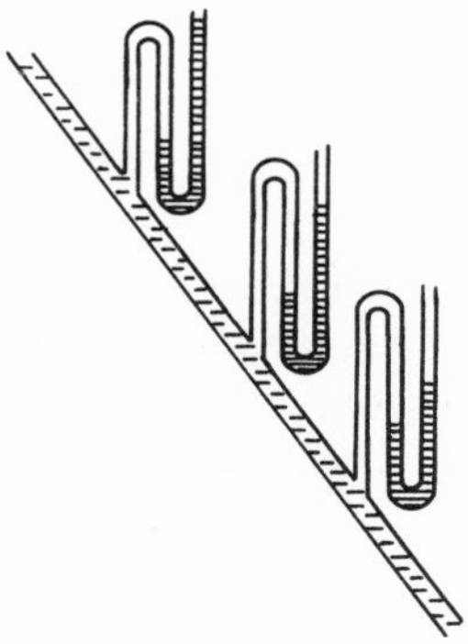

Fig. 1.4. Water Pressure

The establishment of equilibrium is a necessary condition for the determination of a force. There is only a single other quantity that requires equilibrium for measurement, namely, the temperature. Though temperature is conventionally not called a generalized force, it has several characteristics in common with forces.

1.27 Interaction by Contact. Each of the various modes of interaction discussed so far requires a specific interaction gadget: hook and eye for the spring-weight connection, a tube and piston for two gases, a pair of copper wires and a switch for electrical interaction, and so on. There is one kind of interaction left that does not depend on a special gadget but takes place whenever two objects touch each other. This interaction by contact is called thermal interaction. It is prevented only if the objects are separated by an evacuated space. A crude realization of thermal isolation is achieved by a thermos bottle.

It should be understood quite clearly that contact always results in thermal interaction and that thermal interaction can be brought about only by contact.

Insulation by felt, slag wool or similar materials lowers the rate of thermal interaction but it does not isolate the enclosed object. Such materials even rob the enclosed thing of the privilege of being an object because it is in contact with them and therefore cannot be isolated.

On the other hand, thermal interaction can be achieved only by direct contact. Radiation can have similar effects under certain conditions, namely, if reflection, scattering, photoelectric and photochemical reactions are avoided. But the same is true for any kind of interaction. An object can be heated mechanically by friction, electrically by a heating coil and so on; yet these kinds of interaction can never be called thermal.

If an object is thermally insulated from its environment, any change of it is called adiabatic. An "adiabatic change" is, of course, an idealized term of the kind discussed in Section 1.23.

In addition to direct perception of warmer or cooler, the variation of many properties of an object can be used to observe a thermal change. The pressure exerted by a gas enclosed in a rigid vessel is a perfectly satisfactory indication of warmer or cooler, and has been actually used in measurements. The electrical conductivity of a platinum wire and numerous other properties can be used in a similar manner.

We define temperature now just as we defined a generalized force by means of relations (1.9) to (1.11). If two objects A and B are in contact, the one that warms up has the lower temperature, the cooling one the higher temperature. If neither warms up or cools down, they have the same temperature, or are in thermal equilibrium. The definition must be completed by conventions regarding thermometer calibration. The final rules can be discussed only later but the practical scale (volume of a certain amount of mercury indicates 0° at the ice point and 100° at the boiling point of water) presents a perfectly good definition though other definitions will be shown to have practical advantages.

The analogy between temperature and generalized forces is obvious. The fundamental difference is the absence of a coordinate conjugated to temperature. This matter, too, can be discussed only later.

1.28 Work. As in the discussion of forces (Section 1.26), we start with our everyday idea of work, namely, as "the price we pay"(Perrin[14]) for changing the state of an object. The momentary effort is measured by the force required to bring about a change; accordingly, the total effort necessary to convert an object from an initial state I to a final state F is defined as the work done upon it

$$w' = -\int_I^F f''dx''. \tag{1.12}$$

The distinction of the superscripts is important. Since the interaction condition is, as a rule though not always, given in the form of (1.1), we usually may introduce the coordinate change dx' of the object under consideration for the quantity -dx" of the environment.

But the force f", exerted by the environment (i.e., by ourselves or an object serving as our proxy or tool), need not be equal to f', the responding force of the object to be changed. We charge a storage cell at a voltage of 2.4 v; this is the external generalized force f". But the cell responds with a lower force, say 2.0 v, when we discharge it; this is its internal generalized force f'. Only in equilibrium, i.e., with no noticeable current in either direction, the internal and external forces become equal.

If the whole change from I to F may be considered to be a sequence of equilibrium states, the process is called quasistatic or reversible, and the work exerted on the object is given by

$$w' = \int_I^F f'dx'. \tag{1.13}$$

But for irreversible processes we must go back to (1.12).

For the measurement of the work w' done upon an object, it is obviously necessary that the initial and final states of the object are defined; in addition, the force f" exerted by the environment as a function of its coordinate x" must be defined. It is not required that the object's force f' during the process is defined.

According to (1.12), the definition of work is based on the prior concepts of forces and coordinates. This order of defining cannot be reversed. Work is the last of the basic concepts.

A system consisting of two identical objects requires twice the amount of work to be changed from I to F; work is therefore an extensive quantity as defined in the following section.

1.3 THE COMPOSITION OF AN OBJECT

1.31 Extensive and Intensive Properties. A distinction of little theoretical but considerable practical importance results from the fact that our interest is restricted to two classes of problems: total properties of an object, or qualities of a homogeneous object regardless of its amount. If we design, e.g., a plant producing per day 1000 tons of sulfuric acid, we are interested in the total daily amount of sulfur dioxide passing through the stack into the atmosphere. We are also interested, regardless of the amount, in the concentration of sulfur dioxide in the stack gas (moles sulfur dioxide per total number of moles). Both quantities are important pollution indicators.

For this reason we define two classes of properties[15]. We consider two identical objects, let us say two kilogram weights of brass, or two paper cups filled with coffee, or two adsorption towers filled with alumina, and combine the two objects into a single object or a system, without letting them interact in any manner.

Then the first class of properties, which we call extensive, are those whose value for the system is twice the value for each of the objects. Such properties are the volume, or the mass, or the volume of the coffee in the cups, or the weight of alumina in the towers, or the surface area of alumina in the towers.

The second class of interest, intensive properties, are those that have the same value for the two objects and the system: density, dielectric constant, pressure, and so on.

The two classes, of course, are very far from exhausting all possible properties of an object. But we are really not interested in such a property as the square root of the volume, which does not belong to either class.

It is not correct to say that extensive quantities are proportional to the amount or the mass of an object. If we pour mercury from a beaker on a tray, the amount of mercury, however we measure it, is not changed. But the surface area becomes much larger: it is a property independent of the amount. But it satisfies the definition of an extensive quantity. Another example is the electric charge of capacitors in a parallel arrangement.

The classification extensive-intensive has nothing to do with the distinction of generalized coordinates and forces [16].

1.32 Phases. We call a phase or a homogeneous object the total material of an object that has the same intensive properties. A martini on the rocks has two phases (when we disregard the olive), but after the ice has melted it has only one phase.

Even such a plain concept as "phase" leads to difficulties if we wish to satisfy a legalistic mind. We are inclined to call the atmosphere a single phase, but its composition and properties are quite different at an altitude of 10 km or 100 km. There is little gained in arguing this point but if necessary we should be prepared to call the atmosphere a continuous sequence of phases.

A system of two or more phases is called heterogeneous.

1.33 Composition Variables. By components of an object we mean a set of substances necessary and sufficient to build up the object. We use the term in the sense of "independent components." A system containing hydrogen, oxygen and water has, in general, three components. But if mercury vapor or palladium black is present, the formation of water can proceed rapidly; as a result, only water and either hydrogen or oxygen (whichever is present in excess) are sufficient to build up the system. For the count of components the molecular state is irrelevant. Ethanol exists as a monomer in the gas phase, in the form of associated molecules of various sizes in the liquid or in a solution in hexane. Nevertheless it counts as a single component as long as the association proceeds practically instantaneously so that a single molecular state distribution exists for any solution.

We express the amount of a component j in a phase as the number of moles n_j present. The total number of moles in the phase is

$$n = n_1 + n_2 + \ldots + n_z = \Sigma n_j \ . \tag{1.14}$$

The numbers of moles are independent variables. Moreover, it is supposed that each of them can be kept constant, e.g., by means of a semi-permeable membrane. Therefore they are generalized coordinates.

We can also describe the composition of the phase by intensive variables, such as the mole fractions

$$y_1 = n_1/n \ ; \qquad y_2 = n_2/n \ ; \qquad \ldots \qquad y_z = n_z/n \ . \tag{1.15}$$

Because of (1.14) the sum of the mole fractions

$$\Sigma y_j = y_1 + y_2 + \ldots + y_z = 1 \tag{1.16}$$

is unity, i.e., the composition is described by $z - 1$ independent intensive variables. The set of z numbers of moles n_j describes not only the composition but also the amount of the phase. It contains one necessarily extensive variable while the remaining $z - 1$ variables may be extensive or intensive.

In the following discussion of the behavior of properties dependent on the composition variables, we shall keep constant any other independent extensive or intensive variable. The results will therefore be strictly valid only for constant surface areas, constant pressure (if introduced as an independent variable), and so on. In this section and the two following ones we shall call G an arbitrary extensive property so that the conclusions will be generally valid for any (extensive) property.

If we remove one mole from a pure liquid phase containing n moles (e.g., if we remove one mole of water from a liquid phase by vaporization) the extensive property G of the phase decreases by G/n. If the phase is a solution, its composition changes during the removal and the change of G for the removal of a small amount dn_j of the component j is expressed by

$$-\frac{\partial G}{\partial n_j}\, dn_j \ . \tag{1.17}$$

The partial derivation requires constancy of the number of moles of all components except j, and constancy of all other independent variables taken into account in the problem considered. We call the derivative the "partial molal value of the property G for the component j" and write

$$\overline{G}_j = \partial G/\partial n_j \tag{1.18}$$

with the implied understanding of which variables are kept constant.

Any molal quantity is of course intensive and thus independent of the amount of the phase. We can, therefore, transform the definition (1.18) into an expression containing only intensive variables. This is desirable for practical reasons. We solve this problem first for a binary solution, and then for a multicomponent solution.

The discussion of composition in this section and particularly in the two following sections will be used only much later (Chapter 5). It is presented here rather than later so that its "pre-thermodynamic" character is clear. Indeed, the discussion is based just on the distinction between extensive and intensive properties. The reader may postpone the detailed study.

1.34 Partial Molal Quantities for a Binary Solution. The composition of a binary solution is characterized by two independent variables. One of them is necessarily extensive; we choose the total number of moles

$$n = n_1 + n_2. \tag{1.19}$$

As the second one we take

$$y_1 = 1 - y_2. \tag{1.20}$$

We introduce the molal quantity

$$\bar{G} = G/n \tag{1.21}$$

which, as an intensive quantity, depends only on y_1, and express (1.18) as

$$\bar{G}_1 = \left(\frac{\partial G}{\partial n_1}\right)_{n_2} = \left(\frac{\partial(n\bar{G})}{\partial n_1}\right)_{n_2} = \bar{G}\left(\frac{\partial n}{\partial n_1}\right)_{n_2} + n\frac{d\bar{G}}{dy_1}\cdot\left(\frac{\partial y_1}{\partial n_1}\right)_{n_2}. \tag{1.22}$$

From (1.19) and (1.15) we derive

$$(\partial n/\partial n_1)_{n_2} = 1 \tag{1.23}$$

$$(\partial y_1/\partial n_1)_{n_2} = 1/n - n_1/n^2 = (1 - y_1)/n = y_2/n. \tag{1.24}$$

The partial molal quantity is therefore expressed by intensive variables as

$$\bar{G}_1 = \bar{G} + y_2 \cdot d\bar{G}/dy_1. \tag{1.25}$$

Since we have not introduced any preference for subscript 1 over 2, we may exchange the subscripts so that

$$\bar{G}_2 = \bar{G} + y_1 \cdot d\bar{G}/dy_2 \tag{1.26}$$

or, in view of (1.20),

$$\bar{G}_2 = \bar{G} - y_1 \cdot d\bar{G}/dy_1 \tag{1.27}$$

if we return to y_1 as the independent variable.

We multiply (1.25) by y_1, (1.27) by y_2, and add the results. Thus we obtain

$$\bar{G} = y_1 \bar{G}_1 + y_2 \bar{G}_2. \tag{1.28}$$

Subtraction of (1.27) from (1.25) leads, in view of (1.20), to

$$d\bar{G}/dy_1 = \bar{G}_1 - \bar{G}_2. \tag{1.29}$$

The differentiation of (1.28) with respect to the single independent variable y_1 results in

$$\frac{d\bar{G}}{dy_1} = \bar{G}_1 - \bar{G}_2 + y_1 \frac{d\bar{G}_1}{dy_1} + y_2 \frac{d\bar{G}_2}{dy_1} \tag{1.30}$$

or, according to (1.29), in

$$y_1 \frac{d\bar{G}_1}{dy_1} + y_2 \frac{d\bar{G}_2}{dy_1} = 0. \tag{1.31}$$

This famous and useful relation has been independently proposed by Gibbs about 1876, by Duhem in 1886, and by Margules in 1896.

The relations (1.25) to (1.31) exhaust what there is to say about the concentration dependence of properties of a binary (gaseous, liquid, or solid) solution phase.

1.35 Multicomponent Solutions. The relations derived in the preceding section can be generalized without difficulty for solutions of more than two components. They are indispensable for thermodynamic discussions of solutions. But this Section may be skipped by the student interested only in general thermodynamics.

Only z - 1 of the mole fractions y_j are independent in view of relation (1.16). We shall not make use of (1.16) in the following discussion though we may use it in application of the results at any point to eliminate one of the variables y_j. The advantage of this procedure is the maintenance of symmetry of all results in the variables y_j; we mention the disadvantage that a single derivative

$$\left(\frac{\partial \bar{G}}{\partial y_j}\right)_{y_k} ; \qquad k = 1, 2 \ldots j-1, j+1 \ldots z \tag{1.32}$$

has no physical meaning because $\bar{G}$ can be formally represented in different manners (not containing y_z, or y_{z-1}, and so on). But the results, containing only differences of derivatives of the kind (1.32), are not impaired.

The transformation from mole numbers to mole fractions is now given by

$$\partial y_a/\partial n_b = -n_a/n^2 = -y_a/n; \qquad a \neq b \tag{1.33}$$

$$\partial y_a/\partial n_a = 1/n - n_a/n^2 = (1 - y_a)/n. \tag{1.34}$$

As in (1.22) we find for the partial molal quantity of component j

$$\bar{G}_j = \left(\frac{\partial G}{\partial n_j}\right)_{n_k} = \left(\frac{\partial(n\bar{G})}{\partial n_j}\right)_{n_k} = \bar{G} + n\left(\frac{\partial \bar{G}}{\partial n_j}\right)_{n_k} . \qquad (1.35)$$

Since $\bar{G}$ depends on n_j through all y_a's, we have (with $k \neq j$)

$$\left(\frac{\partial \bar{G}}{\partial n_j}\right)_{n_k} = \frac{1}{n}\sum_a \frac{\partial \bar{G}}{\partial y_a} \cdot \frac{\partial y_a}{\partial n_j} = \frac{1}{n}\left[(1 - y_j)\frac{\partial \bar{G}}{\partial y_j} - \sum_k y_k \frac{\partial \bar{G}}{\partial y_k}\right] . \qquad (1.36)$$

The first term in the brackets takes care of the case a = j, the second of all others. We introduce (1.36) into (1.35) and let a again assume all values including j. Thus we obtain the result

$$\bar{G}_j = \bar{G} + \frac{\partial \bar{G}}{\partial y_j} - \sum_a y_a \frac{\partial \bar{G}}{\partial y_a} . \qquad (1.37)$$

The equation is valid whether or not one of the mole fractions has been eliminated in G.

As in (1.28) we multiply (1.37) by y_j and obtain by taking the sum over all values of j

$$\bar{G} = \sum_j y_j \bar{G}_j . \qquad (1.38)$$

Also from (1.37) follows the difference

$$\bar{G}_a - \bar{G}_b = \partial\bar{G}/\partial y_a - \partial\bar{G}/\partial y_b , \qquad (1.39)$$

which, by the way, has a physical significance.

Multiplying (1.38) by n we obtain

$$G = \sum_a n_a \bar{G}_a$$
$$\frac{\partial G}{\partial n_b} = \bar{G}_b + \sum_a n_a \frac{\partial \bar{G}_j}{\partial n_b} . \qquad (1.40)$$

Introducing (1.18), we obtain the relation of Gibbs, Duhem and Margules

$$\sum_a n_a \frac{\partial \bar{G}_a}{\partial n_b} = 0. \qquad (1.41)$$

This relation holds for <u>any</u> arbitrary variation of the composition variables. It holds also for a variation in which the total number of moles is unity so that n_j has the same value as y_j. Therefore we may write also

$$\sum_a y_a \frac{\partial \bar{G}_j}{\partial y_b} = 0 . \qquad (1.42)$$

In the differentiation we have to keep constant all other variables that have been selected to be independent. That includes all mole fractions except y_b.

SUMMARY OF THE FIRST CHAPTER

Thermodynamics provides a basis for all physical sciences. Its own basis must therefore be established wihout recourse to other sciences. Fundamental concepts are those that are indispensable in the description of nature. They are: object, isolation and interaction; generalized coordinates and forces; equilibrium.

The choice of a generalized coordinate for a mode of interaction is determined by the interaction mechanism and the resulting interaction condition. A generalized force can be measured only in equilibrium. It represents, in crude language, the momentary effort required to change the state of an object according to our pleasure. The total effort for changing an object from an initial to a final state is represented by work.

Molal and partial molal properties of solutions, useful in their description, are related with one another by some general relations.

REFERENCES

(11) W. Shakespeare, "King Edward III," Act 1, Scene 2.

(12) A fascinating account of some errors in science has been given by I. Langmuir, General Electric Research and Development Center, Schenectady, N.Y., Report 68-C-035 (1968).

(13) O. Redlich, J. Phys. Chem., 66 (1962) 585.
Revs. Modern Phys., 40 (1968) 556.

(14) J. Perrin, Bull. Soc. Franc. Philosophie, 6, (1906) 81.

(15) R. C. Tolman, Phys. Rev. (2) 9 (1917) 237.
G. N. Lewis and M. Randall, Thermodynamics, McGraw-Hill, New York, 1923.

(16) O. Redlich, J. Chem. Ed., 47 (1970) 154.

CHAPTER 2. THE TWO LAWS OF THERMODYNAMICS

The awe inspiring power of thermodynamics rests above all upon the strict validity and wide applicability of the two laws. The third law, also a statement of fundamental importance, is not quite so definite and is actually part of statistical rather than pure thermodynamics [21].

2.1 THE FIRST LAW; ENERGY

The first law is, in a way, a product of technical endeavor, namely, of the attempts to build a perpetual motion machine. That this is an impossible undertaking, may be considered the fundamental principle underlying the first law.

The following presentation is based on Carathéodory's method [22]. To be meaningful, any statement of the first law must be prepared in advance by the introduction of the concepts object, generalized coordinates and forces, work, and adiabatic process. The first chapter contains this preparation.

At first, we consider the work required for a few examples of adiabatic changes. Joule's famous measurements showed that a mechanical stirrer can increase the temperature of an adiabatically insulated amount of water, and that a work of 424.85 kgf•m (force of 1 kg exerted over a path of 1 m) is required to heat 1 kg of water by 1°C. The same effect is obtained by means of an electric resistor if 4186 volt•ampere•sec = 4186 watt•sec = 4186 j are used up in the resistor. In this observation, 1 kgf•m is thus found to be equivalent to 9.807 j.

As another example we choose 1 mole of helium at 1 atm and 25°C, adiabatically isolated (Fig. 2.1).

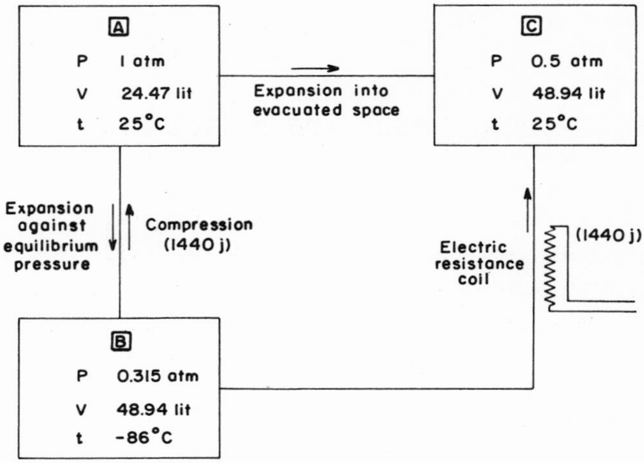

Fig. 2.1. Expansion and Compression of 1 Mole Helium

We can let it expand under various conditions:

(a) If we let one mole of the gas expand from state A (1 at, 24.47 lit) into an evacuated vessel to twice the initial volume, the pressure goes down to 0.5 atm and the temperature remains at 25°C (state C). No work is done by the helium or upon the helium.

(b) If we let the gas expand to twice its initial volume against a movable piston which is kept loaded close to the equilibrium pressure, the pressure goes down to 0.315 atm, the temperature to -86°C (state B). The expanding gas does an amount of work upon the piston which is equal to 147 kgf•m or 1440 j. This type of adiabatic expansion is reversible because the same sequence of equilibrium states can be run through in the reverse order by loading the piston with arbitrarily small additional loads.

(c) We can also keep the volume constant at 48.94 lit and increase the temperature inside the adiabatic enclosure by means of an electric resistor. We arrive at a temperature of 25°C (state C) after putting in 1440 volt• ampere•sec = 1440 j, the same amount that the gas lost in the reversible adiabatic expansion from A to B.

Summarizing we see that the gas can be changed <u>adiabatically</u> from B to C either by way of A or directly. The <u>work required</u> is the <u>same</u> in either case.

As another example we consider the desalting of sea-water. In the osmotic apparatus at the left (Fig. 2.2) seawater and pure water are

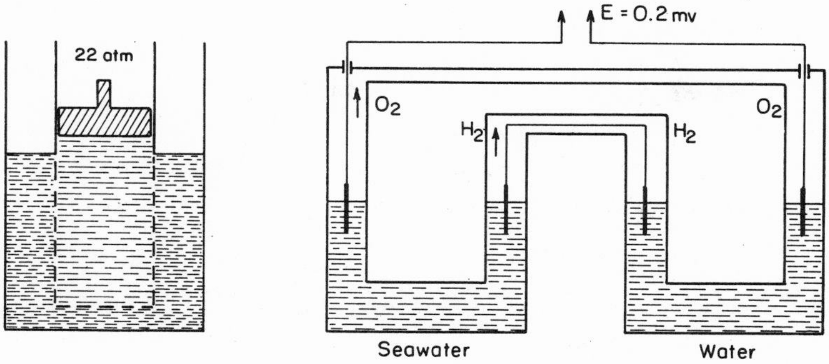

Fig. 2.2. Conversion of Seawater Into Pure Water

separated by a semipermeable membrane which lets water pass through but not the salts. The two phases are in equilibrium if the pressure on the seawater is 21.0 atm higher than the water pressure. Under this pressure head a reversible separation is possible. It requires 21.0•0.01802 = 0.378 lit.atm or 3.91 kgf•m or 38.4 j per mole of water.

The same separation can be brought about by switching two galvanic cells against each other in which water is reversibly decomposed with the aid of electrodes kept in hydrogen and oxygen. The voltage required for decomposition of water in the seawater cell is 0.00020 v higher than the other cell voltage. The difference of the two voltages (times the electric charge of the ions of one mole of water) furnishes the total work required. It is 2•96487•0.00020 = 38.4 j, the same amount as used in the osmotic method [23].

The principle introduced by Carathéodory for the first law generalizes such observations. It states: The amount of work required for adiabatically changing an object from an initial state I to a final state F is always the same, independent of the path chosen, of the nature of the interaction (mechanical, electrical, and so on), and of the other circumstances. The work required depends only on the nature of the object, its initial state, and its final state. It is therefore a state function of the object, called the difference of energy $E'_F - E'_I$ of the object between the two states F and I. It can be measured for an adiabatically enclosed object by (cf. Eq. 1.12)

$$E'_F - E'_I = -\int_I^F \sum_i f''_i dx''_i \tag{2.1}$$

i.e., by means of the forces f''_i and coordinates x''_i of the environment. The only information required regarding the object is the characterization of the initial and final states.

If the change is not only adiabatic but also reversible the energy can be expressed (cf. Eq. 1.13) as

$$E'_F - E'_I = \int_I^F \sum_i f'_i dx'_i \tag{2.2}$$

by properties of the object alone.

For the example of the helium expansion (Fig. 2.1) we conclude that the energy in the states A and C is equal, and that

$$E_A = E_C = E_B + 1440 \text{ j.} \tag{2.3}$$

The energy of a system of non-interacting objects is equal to the sum of the energies of the constituents because the same adiabatic work must be introduced into each constituent whether we consider it alone or as part of a system. Energy is therefore an extensive quantity.

2.2 HEAT

Carathéodory's method enables us now to introduce the concept of heat, or better, of the amount of heat, in an extremely simple manner.

Returning to Fig. 2.1 we see that the helium could be changed from state B to state C without any expense of work since the volume in both states is the same, namely, 48.94 lit. Naturally the change cannot be brought about adiabatically since we must have <u>some</u> agent influencing the object. But if we remove the adiabatic enclosure, we may establish thermal interaction, say with the hot gases from a Bunsen burner, and heat the helium (at constant volume) to 25°C. Thus we obtain an energy increase of 1440 j without doing any work.

We <u>define</u> the heat q' introduced (in a non-adiabatic or <u>diathermal</u> change) into an object as

$$q' = E'_F - E'_I - w' \tag{2.4}$$

i.e., as the energy increase of the object reduced by the work w' done upon the object. Before Carathéodory, this equation was considered to be the expression of the first law. But neither heat nor work were defined in a satisfactory manner.

That this definition coincides with the common idea of heat, can be easily seen. We consider a system of two objects, isolated from the environment. No interaction with respect to any coordinate is established but the two objects are connected by a heat conducting material, e.g., a rod of copper or aluminum. In the initial state the energy of the system E_I is the sum of the energies of the constituents and the same is true for the final energies (Fig. 2.3):

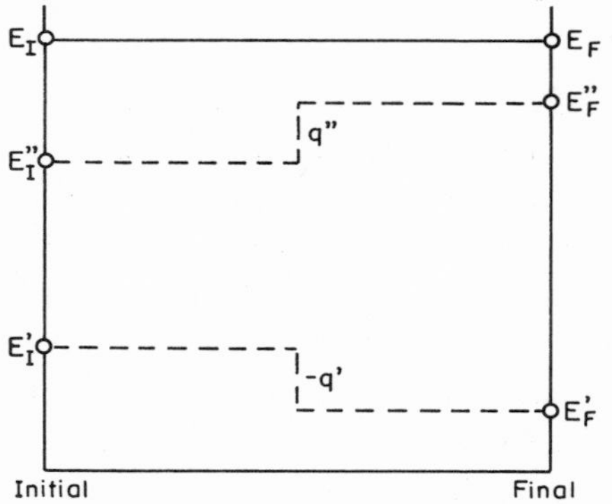

Fig. 2.3. Heat Flow Inside an Isolated System

$$E_I = E'_I + E''_I; \qquad E_F = E'_F + E''_F. \tag{2.5}$$

Since the system is isolated its total energy cannot change so that

$$E_F - E_I = E'_F - E'_I + E''_F - E''_I = 0. \tag{2.6}$$

Therefore we obtain for the heat that flows into the second object

$$q'' = E''_F - E''_I = -(E'_F - E'_I) = -q' \tag{2.7}$$

since no work is done on either object.

In other words, in thermal interaction energy flows from one object to another, and this energy transfer is called heat flow.

Heat is of course an extensive quantity since work and energy are extensive.

2.3 THE SECOND LAW; ENTROPY

Carnot's problem of the maximum efficiency of an engine is closely tied to the question of the direction in which a change proceeds. The most attractive technical process is not useful if the actual reaction runs in the opposite direction. The answer for the simplest case (interaction between two objects with respect to a single coordinate under adiabatic conditions) has been discussed when generalized forces were introduced (Section 1.26). As soon as more than one coordinate is introduced, the problem of the direction of a process requires a much more elaborate consideration.

The second law is based in the following on an idea of Carathéodory [22], modified by Buchdahl [24].

2.31 Accessibility. Before we can discuss the problem of the direction of a process, we have to survey the possible states of an object and their accessibility. In introducing the idea of generalized coordinates, we pointed out that we are able to change any coordinate according to our pleasure by establishing interaction with another object. The number of coordinates (or modes of interaction) is different for different problems but it is a definite number h for a given problem. When all coordinates are fixed we can still change the object by thermal interaction. Thus the possible states constitute an (h+1)-dimensional manifold.

In a general sense, all states of an object are accessible. We can convert a tree with the appropriate amount of oxygen into carbon dioxide and water. The reverse process, the construction of a tree, is more difficult but with good tools such as a greenhouse it is still possible. A completely inaccessible state would not be a state of the object.

An isolated object, however, is severely restricted in the attainment of other states. An isolated gas can expand into a vacuum to attain greater volumes but it cannot attain a smaller volume.

Can we mitigate the severe condition of isolation and yet obtain some significant information of the direction of natural processes? A distinction of accessible and inaccessible states would furnish this information.

Such a distinction certainly cannot be made with respect to any coordinates. We have introduced coordinates as variables that we can change according to our pleasure by establishing interaction of the object under consideration with other objects. Thus any combination of values of h coordinates is always attainable. Any such combination is attainable even under a restriction, namely, the restriction to adiabatic changes.

But there are other states, which can be reached by thermal interaction. If we keep all h coordinates fixed we can still introduce or take out energy by heating or cooling. Can we reach these states also without heating or cooling, i.e., under adiabatic conditions? One can see immediately that the answer is yes for states that are accessible from the initial state by heating. Indeed, we can impose electric work on the object and use it inside the object to increase its energy by means of a resistor. The same effect can be achieved by adiabatic stirring or friction, i.e., by mechanical work.

But the answer is no for states accessible from the initial state by cooling. We cannot (adiabatically) attain any state with the same values of the coordinates and a lower energy. There are in the vicinity of any state others that cannot be reached adiabatically. This fact has been enounced by Carathéodory as the principle on which he was basing the second law. We shall introduce a formulation which is somewhat different from Carathéodory's.

The distinction between (adiabatically) accessible and non-accessible states can be easily shown for an object that depends on a single coordinate. The diagram in Fig. 2.4 represents the energy E as a function of the coordinate x. As an example we may take an adiabatically enclosed storage cell; the charge is then given by x, the voltage because of (2.2) by the slope of the reversible curve F_a.

In a reversible charging operation the state of the cell is represented by the points of the curve above I, in a discharge by the lower points of the curve. But if we short-circuit the cell, it does not do any work so that the energy E remains constant while the charge x decreases. The state point moves therefore into the half-plane F_b. There is no means of increasing the charge without increasing the energy, i.e., the state point cannot move in the half-space F_c.

The diagram shows also a diathermal process at constant charge x. If we increase the energy by heating the cell, the state point moves from I in the direction of the vertical arrow into the half-plane F_b.

For a single coordinate the division of the half-planes F_b and F_c by the reversibility curve F_a is obvious. In this special case we need not introduce a new principle. However, for the general, multidimensional case this division cannot be shown intuitively. Here we introduce the principle of the second law, comprehending the results of multifarious observations: If we change an object starting from an initial state I by adiabatic processes, (a) there is a set of states F_a that can be reached reversibly, (b) there is also a set of states F_b that can be reached irreversibly, and (c) there is a set of states F_c which cannot be reached adiabatically at all.

Now we may consider the diagram of Fig. 2.4 as a symbolic representation of the three sets of states F_a, F_b, and F_c, although the essence of the discussion will be independent of any graphical representation; the diagram serves only to make the discussion easier.

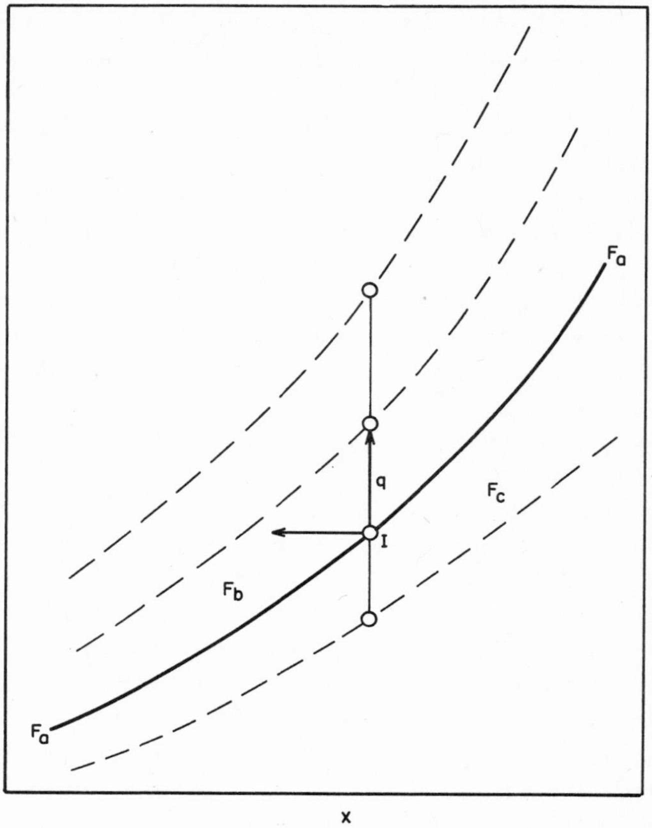

Fig. 2.4. Energy E in the State Space (the curves are isentropic lines; I initial state; the states on the line F_a are from I adiabatically-reversibly accessible; the states in the field F_b are irreversibly accessible; the states in F_c are adiabatically inaccessible)

We select a point I' in the half-space F_b as a second starting point and mark the set of states F_a' which are (adiabatically) reversible from I' (one of the broken lines in Fig. 2.4). No point of F_a' is reversibly accessible from I or F_a; otherwise <u>any</u> point of F_a' (including I') could be reached from <u>any</u> point of F_a (including I), contrary to the selection of I' as a point in F_b. Similarly we select a state I" (that does not lie either in F_a or F_a') and find a set of states F_a'' of (adiabatically) reversibly accessible states, and so on. No two members of the sets F_a, F_a', F_a'' ... have any state in common.

For this reason we can distinguish between the sets F_a, F'_a, F''_a ··· by assigning to them (cf. Section 1.13) different values of a quantity S, which we call the entropy of the object. The calibration of the entropy is essentially arbitrary. We stipulate for the points of class (b), which can be adiabatically-irreversibly reached from I, that S_b is greater than S_I. This means that for class (c) the entropy S_c must be smaller than S_I since I is irreversibly accessible from any point of class (c). The border between the classes (b) and (c) is the class (a) of all points reversibly accessible from I; this is expressed by constancy of S_a. Thus the classification of states is given by

(a) $S_a = S_I$ or $dS = 0$ (adiabatic-reversible)

(b) $S_b > S_I$ or $dS > 0$ (adiabatic-irreversible)

(c) $S_c < S_I$ (no adiabatic process).

The introduction of the concept of generalized forces in Section 1.26 enabled us to describe the interaction of two objects with respect to a single coordinate. The problem of describing the behavior of an object depending on two or more coordinates requires the introduction of a new state function, the entropy.

2.32 Entropy and Temperature. There is still the problem left of assigning numerical values and describing the measurement of entropy. For this purpose we keep all coordinates x constant and increase (or decrease) the energy of the object by introducing (or withdrawing) heat. This change is represented in Fig. 2.4 by the vertical straight line through the point I. We select a few of the points of this vertical line (marked by circles in the diagram) and connect each point with all points in the state space that can be reached adiabatically-reversibly. We find these points experimentally. This adiabatic-reversible set (curve for a two-dimensional state space, surface for a three-dimensional state, hypersurface for a multi-dimensional space) is called an isentropic. A family of isentropics is indicated in Fig. 2.4 by a few broken curves. Two isentropics cannot intersect because in this case each point of the first would be (adiabatically-reversibly) accessible from each point of the second, and thus the two isentropics would contain only points of a single isentropic.

In the assignment of numbers to the entropy we are restricted only in that the same number must be assigned to all points of an isentropic, and that the function S must be monotonically increasing in the direction of an irreversible process. Figure 2.4 indicates also that it must increase if the energy is increased at constant coordinates, i.e., by introduction of heat. However, we cannot set the entropy increase from one isentropic to another equal to the heat introduced because the energy increase varies when we go from one value of x to another one. We settle (actually we postpone) the problem by defining a new state function θ by the relation

$$dS = \theta dE \tag{2.8}$$

and have to show how to measure θ. Since both E and S are functions of the state, so is θ.

We define S as an extensive quantity. The function θ therefore must be intensive, and for a system of non-interacting objects S is additive

$$S = S' + S'' \tag{2.9}$$

as E is (see Eq. 2.5).

To find the meaning of θ we consider an isolated system consisting of two objects. All coordinates are keptconstant so that only heat flow from one object to the other can take place.

The heat flow is therefore controlled by the following relations:

$$dE' + dE'' = 0; \quad dE'' = -dE' \tag{2.10}$$

$$dS' = \theta' dE'; \quad dS'' = \theta'' dE'' = -\theta'' dE' \tag{2.11}$$

reversible: irreversible:

$$dS = 0; \qquad dS > 0 \tag{2.12}$$

$$dS = (\theta' - \theta'')dE' = 0 \qquad dS = (\theta' - \theta'')dE' > 0. \tag{2.13}$$

For the irreversible case we have either

$$\theta' > \theta''; \qquad dE' > 0 \tag{2.14}$$

or

$$\theta' < \theta''; \qquad dE' < 0. \tag{2.15}$$

Therefore heat flows _to_ the first object i its value θ' is greater, and conversely. If

$$\theta' = \theta'' \tag{2.16}$$

we have either equilibrium (dE' = 0) or a reversible heat flow, i.e., a sequence of equilibrium states which is practically realized if (2.16) is approximately satisfied. The discussion is obviously analog to the discussion of the three relations (1.9) to (1.11) for generalized forces.

The quantity θ therefore controls the flow of heat. But the discussion of thermal interaction in Section 1.27 has shown that the property exerting this control is the temperature. It is important to see that <u>both</u> θ and T control the heat flow whatever else happens, i.e., whatever other properties the two objects have. For this reason θ must be a universal function of the temperature, independent of any other variable. In other words, θ and T express the same property, with the only difference that heat flows to the object with the <u>higher</u> value of θ and the <u>lower</u> value of T. This fact restricts our calibration either of θ or of T. Otherwise we are still free to calibrate both variables. We choose the simplest permissible relation, namely,

$$T = 1/\theta \tag{2.17}$$

so that (2.8) requires

$$dS = dE/T. \tag{2.18}$$

In this equation the change dE signifies a change at <u>constant coordinates</u>.

Equation (2.18) is usually written with the quantities q_{rev} and w_{rev} of a reversible change, namely,

$$dS = q_{rev}/T = (dE - w_{rev})/T. \tag{2.19}$$

This is correct but a little too narrow. Indeed dE at constant coordinates is an amount of heat. The measurement of entropy, however, does not require that the change is reversible but only that the temperature of the object (and the work done by it) are defined; this does not imply that the environment is in equilibrium. Actually we measure entropy changes (together with the heat capacity) in a calorimeter by means of irreversible processes. Only the <u>temperature</u> of the <u>calorimeter</u> must be well defined. For the measurement of <u>work</u>, on the contrary, we need the <u>coordinates</u> and <u>forces</u> of the <u>environment</u>.

The entropy is defined by (2.18) as an integral function, i.e., only the change of entropy between two states of an object can be measured. Equation (2.18) shows that a zero point of temperature has no meaning in pure thermodynamics. Zero points of both entropy and temperature, however, are introduced in statistical thermodynamics, and are important and useful.

2.33 <u>The Calibration of Temperature; Heat Engines.</u> So far we have defined temperature only qualitatively. According to Section 1.13, we have to complete the definition by assigning numbers to the different temperatures. In principle, the choice is arbitrary within a group of monotonically increasing functions. Actually Kelvin once proposed to use log T in the place of T. The suggestion was not accepted because the undeniable advantages of this choice were outweighed by disadvantages. But the proposition was by no means unreasonable.

The choice is important because the temperature is involved in so many thermodynamic relations. In particular, the calibration of the entropy depends on that of T, according to (2.19). As a matter of fact, we can use this relation for the calibration of the temperature by means of the idealized heat engine (Carnot's cycle).

This cycle consists of four reversible steps. In the first, the heat q is put into the engine at the constant temperature T from a heat reservoir. The engine temperature is adiabatically lowered to the temperature T_o in the second step. In the third, the engine gives off the heat q_o (or receives the heat $-q_o$) at the temperature T_o. In the last step it reverts adiabatically to the initial temperature T. The engine does work in the first two steps and receives work in the third and fourth. If we choose the proper ratio q_o/q, the engine reverts to its initial state. To do this it is necessary that the total changes of the engine in energy as well as in entropy are zero. This is also sufficient because the state of the engine is determined by two independent variables. Since the adiabatic-reversible steps do not change the entropy, we have

$$\frac{q}{T} - \frac{q_o}{T_o} = 0. \tag{2.20}$$

The energy is balanced by the work -w done by the engine (w work done upon the engine) in one cycle

$$q - q_o - w = 0. \tag{2.21}$$

Carnot's result was therefore that the maximum efficiency in the conversion of heat into work is

$$w/q = (q - q_o)/q = 1 - q_o/q = 1 - T_o/T. \tag{2.22}$$

The important conclusion was that the maximum conversion efficiency depends only on the ratio of the two temperatures (for a steam engine the temperature of the boiler and the condenser).

For our problem it follows that we can quantitatively define the <u>thermodynamic</u> or <u>absolute</u> <u>temperature</u> from the efficiency of a Carnot cycle. Since from (2.22)

$$T_o/T = 1 - w/q \quad \text{or} \quad T = T_o q/(q - w) \tag{2.23}$$

we can assign a numerical value to T as soon as we have arbitrarily fixed a single temperature T_o; the ratio $q/(q - w)$ can be determined experimentally. The present convention for T_o is

$$T_o = 273.16 \text{ K} \qquad (2.24)$$

for the triple point of water, i.e., for the temperature at which ice, water and steam are in equilibrium. (The designation K means "degrees Kelvin"). This temperature can be easily reproduced within 0.01° and without excessive effort within 0.0001°.

The practical standardization of temperatures is based on relations involving quantities that can be more accurately measured than w/q. But the principle is the same since any thermodynamic relation containing T must lead to the same calibration.

2.4 CHARACTERISTIC FUNCTIONS

The standard problem of chemical thermodynamics can be expressed in the following manner. First we measure for each pure substance several thermodynamic properties and tabulate them or express them algebraically as functions of the temperature and pressure. Secondly, we use these functions to predict whether a prescribed reaction between some substances proceeds irreversibly, or is impossible, or leads to an equilibrium.

The first step requires a large amount of experimental information. The magnitude of the task is considerably reduced by the use of theoretical relations, particularly regarding dependence of some functions on the temperature and pressure.

But the application of the data in the second step furnishes a still very much larger amount of practical information since the number of possible reactions is much larger than that of the reaction partners. For any reaction we want to know the conditions of equilibrium, the amount of work possibly obtained by means of the process, and various information regarding dependence on temperature and pressure.

If a solution, i.e., a phase containing several components, participates in the reaction, the data problem as well as the information gain become much larger because of the influence of the composition of the phase. We shall at first consider pure phases or solutions whose composition remains unchanged. Later on solutions will be generally discussed.

The behavior of the entropy in an adiabatically enclosed system furnishes us some hint of how the standard problem is to be solved. The classification of processes according to the three cases $dS = 0$, or $dS > 0$, or

$dS < 0$ contains the answer. But the entropy is a suitable decision function only for isolated or at least adiabatically insulated systems. We shall see that other functions are more suitable for the most frequent practical conditions, i.e., for isothermal reactions, and for reactions proceeding at constant temperature and pressure.

The principal functions which we shall discuss are "characteristic functions" (they are not the only ones in this class). The name was coined by Massieu a century ago. A characteristic function, expressed as a function of two suitable variables, furnishes all other thermodynamic properties of interest by differentiation (and elementary operations) without an integration. The distinction is important because no integration constant is introduced in these operations. In other words, the characteristic function and its independent variables must already contain the two integration constants for the energy and the entropy.

The following discussion is inevitably somewhat abstract. A proper familiarity with the functions and relations discussed can be gained only by practical use in solving problems and in other applications. In many presentations only the reversible work $-PdV$ done by the external pressure is taken into consideration. But x_i in the following discussion stands also for the surface area or the electric charge or any other generalized coordinate, and f_i stands for the conjugate force. If we leave out the corresponding contributions to the work, we have to patch them on later -- a procedure that has often seriously weakened the power of thermodynamic discussion.

2.41 Entropy. We consider the energy change dE of an object for a change from a certain initial to a certain final state. We keep the two states unchanged and compare a reversible and an irreversible transition. For the reversible work we may write

$$w_{rev} = \sum_i f_i dx_i = \Sigma f dx. \qquad (2.25)$$

The energy change is the same for the two transitions

$$dE = q + w = q_{rev} + \Sigma f dx. \qquad (2.26)$$

If we go from the chosen initial state to the chosen final state in a reversible process we have, according to (2.19)

$$dS = q_{rev}/T. \qquad (2.27)$$

In an irreversible process from the same initial to the same final state, dS must be the same, being the difference of a property between two states. We split now an irreversible process into two steps. In the first step we introduce reversibly the heat q so that the entropy change is q/T; the second step proceeds irreversibly so that the entropy necessarily increases by a positive number which we call η/T. For the total process we have

$$dS = (q + \eta)/T. \tag{2.28}$$

If we succeed to reach the same final state the last two equations furnish

$$\eta = q_{rev} - q > 0. \tag{2.29}$$

This means that the heat q introduced into the object in the irreversible process must be (algebraically) smaller than the heat introduced in a reversible transition from the same initial to the same final state. If this condition is not satisfied we cannot reach the prescribed final state. We may consider η as a measure of the irreversibility of a process. Equilibrium (or a reversible change) is bound to the condition $\eta = 0$.

From (2.26) we conclude

$$w = q_{rev} - q + \Sigma f dx = \eta + \Sigma f dx. \tag{2.30}$$

The work we have to do for an irreversible process with prescribed initial and final states is greater than the work for the reversible process.

2.42 Free Energy of Helmholtz. The free energy of Helmholtz, also called maximum work, is defined by

$$A = E - TS. \tag{2.31}$$

From (2.26) and (2.27) one concludes

$$dE = q_{rev} + \Sigma f dx = TdS + \Sigma f dx \tag{2.32}$$

so that the differentiation of (2.31) furnishes

$$dA = -SdT + \Sigma f dx = -SdT + w - \eta. \tag{2.33}$$

This function is useful for isothermal changes since its increase is equal to the work reversibly done upon the object at constant temperature, i.e., the minimum work that must be done upon the object to change it from a given initial state to a give final state.

If the object is kept in a thermostat under such conditions that it cannot do any work (closed vessel), the Helmholtz free energy decreases according to (2.33) in an irreversible process because under these conditions we have $dT = 0$; $w = 0$, and η has always a positive value.

2.43 Free Energy of Gibbs. Chemical changes proceed often not only at constant temperature but also at constant pressure. For this case a new set of functions is most suitable. We introduce the enthalpy or the heat content

$$H = E + PV \tag{2.34}$$

and the free enthalpy or the free energy of Gibbs

$$G = H - TS = A + PV = E - TS + PV. \tag{2.35}$$

This function has been called thermodynamic potential by Gibbs [25].

In addition, we introduce the useful or net work $\bar{w}$, which is the total work except for the work $-P\Delta V$ done on the object by the pressure

$$\bar{w} = w + PdV. \tag{2.36}$$

For the reversible net work we write

$$\bar{w}_{rev} = \sum_i fdx + PdV = \Sigma' f_i dx_i = \Sigma' fdx \tag{2.37}$$

indicating by the prime that the volume work is to be left out in the summation. The reason for introducing the net work is, of course, that we wish to apply the functions of this section to problems where the work of the outside pressure is always done without being of any particular interest. If no gases participate in the process, the volume work is usually small, sometimes negligible. Historically, this fact has contributed to some confusion of the functions of Helmholtz and Gibbs.

From (2.33) and (2.35) to (2.37) we obtain

$$\begin{aligned} dG &= dA + PdV + VdP \\ &= -SdT + \Sigma fdx + PdV + VdP = -SdT + VdP + \Sigma' fdx \\ &= -SdT + w + PdV + VdP - \eta = -SdT + VdP + \bar{w} - \eta. \end{aligned} \tag{2.38}$$

For an isothermal-isobaric change we have

$$dG = \Sigma' fdx = \bar{w} - \eta. \tag{2.39}$$

This is the most frequently used basic relation of chemical thermodynamics. It means that in a reversible isothermal-isobaric change the Gibbs free energy increases by the useful work done upon the system; in an irreversible process a greater amount of useful work must be done.

Frequently one finds the relation

$$\Delta G \lesseqgtr 0. \tag{2.40}$$

This is of course correct for an isothermal-isobaric change if no useful work is done. But if we charge a storage cell (Fig. 2.5), ΔG has a

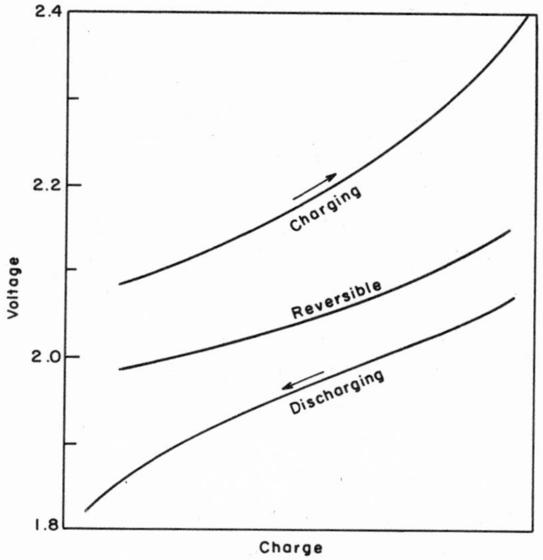

Fig. 2.5. Lead Storage Cell (ΔG is indicated by the area under the reversible or "open circuit" voltage; the work required for charging and the work obtained by discharging are given by the areas under the other curves. The reversible voltage changes with the concentration of the sulfuric acid).

considerable positive value because the electrical work (voltage times charge) is introduced into the cell. We see immediately the influence of the irreversibility from (2.39). In charging, $\bar{w}$ is positive and the work required for a certain dG exceeds the reversible charging work by the amount η. But in the discharge dG has the same value with a negative sign, and the amount of (negative) work obtained is smaller by η than the reversible work. For this reason the charge voltage is higher, and the discharge voltage lower than the reversible voltage, which can be measured by means of a potentiometer practically without a current in either direction.

The practical application of (2.39) or (2.40) is based on our knowledge of the free energy values of a very large number of various substances, and on knowledge of the variation of the free energy as a function of temperature, pressure, and, in solutions, of the concentrations of the constituents. This knowledge stems from extensive measurements as well as elaborate theories.

SUMMARY OF THE SECOND CHAPTER

According to the first law, in Carathéodory's formulation, we have always to spend the same amount of work when we adiabatically change an object from a certain initial to a certain final state. This work is called the increase of the energy. Heat is the difference between the energy increase and the work spent in any specific (diathermal) change.

Experimentally we find that, starting from a certain state of an object, we can divide all possible states in three classes: (a) Some states are reversibly accessible in an adiabatic change. (b) Some states are adiabatically-irreversibly accessible. (c) Some states are not adiabatically accessible. These facts are a sufficient basis for a definition of the entropy S as the property that is unchanged in a change of class (a), and increases in class (b).

Heat flow is linked to entropy change and temperature. The calibrations of entropy change and temperature are connected.

For quantitative applications the most useful thermodynamic function is the thermodynamic potential or the free energy of Gibbs, represented as a function of temperature and pressure. In a reversible isothermal-isobaric change the decrease of the Gibbs free energy is equal to the useful or net work done by the object (work done except against the external pressure). An example is the electric work done by a storage cell.

REFERENCES

(21) A statement called "zeroth law of thermodynamics" in recent times does not need any discussion in this place.

(22) C. Carathéodory, Math. Ann., 67 (1909) 355.

(23) No practical desalination process could be based on this method because of the very high current required.

(24) H. A. Buchdahl, Am. J. Phys., 17 (1949), 41, 44, 212; 22 (1954) 182; 23 (1955) 65. Z. Physik, 152 (1958) 425. Buchdahl later was dissatisfied with his first version and found it necessary, for mathematical reasons, to present a considerably more complicated version [H. A. Buchdahl and W. Greve, Z. Physik, 168 (1962) 386]. In this book, on the contrary, a greatly simplified version of the original idea is presented.

(25) Helmholtz introduced the function A under the symbol F. Early in this century it was widely and negligently used where the thermodynamic

potential would have been correct. For this reason Lewis and Randall (Thermodynamics, McGraw-Hill, New York, 1923) changed the names and symbols: They called the thermodynamic potential the "free energy (F)" and Helmholtz's function "maximum work (A)". This practice was widely adopted, but in recent years the names and symbols used in this book have been generally preferred.

CHAPTER 3. A SINGLE PURE PHASE

The influence of temperature and pressure on the thermodynamic functions will be examined in this chapter.

For this purpose, the discussion will be restricted in two ways. The object under consideration will be a single phase, and the only coordinate taken into account will be the volume. Therefore the only work considered will be the work done by the pressure, and the useful work done will be zero. This does not mean that other coordinates do not exist; their influence will be superimposed on the results we shall obtain in this chapter. But right at the start we want to avoid the impression that the discussion here is fully general in the same sense as it was in the earlier chapters. The results will be generally applicable but they will often constitute only part of a full description.

In particular, they will be applicable to a solution but only as long as no change of the composition is involved in the problem.

The discussion will develop from the relations (2.32), (2.33), and (2.38) which in this chapter will appear as

$$dE = TdS - PdV \tag{3.1}$$

$$dA = -SdT - PdV \tag{3.2}$$

$$dG = -SdT + VdP. \tag{3.3}$$

Since we admit only two modes of interaction, thermal and volume, there are only two independent variables. For very good reasons, T and P rather than S and V are almost always preferred for practical applications. One reason is the fact that they are in general preset by practical considerations. Even more important is the fact that in an equilibrium of several phases they usually have the same value throughout the system; the entropies and volumes of the various phases are different from one another.

Under the restrictions of this chapter, any extensive property is easily converted into an intensive property, namely, into a molal quantity through division by the number of moles, or into a specific quantity through division by the weight of the phase. This is not so for a system of several phases, or whenever other coordinates (interface area, electric charge, and so on) have been introduced.

3.1 DEPENDENCE ON THE TEMPERATURE

Equations 3.2 and 3.3 show us directly the temperature dependence of the two free energies A and G, namely,

$$\left(\frac{\partial A}{\partial T}\right)_V = -S ; \qquad \left(\frac{\partial G}{\partial T}\right)_P = -S. \tag{3.4}$$

Here and in the following differentiations all other coordinates besides V are to be kept constant.

Some earlier authors preferred to eliminate the entropy by combining (2.31) and (2.35) with (3.4) so that

$$\begin{aligned} E &= A + TS = A - T(\partial A/\partial T)_V; \\ H &= G + TS = G - T(\partial G/\partial T)_P. \end{aligned} \tag{3.5}$$

Calorimetric measurements provide information of the change of energy ΔE or the change of heat content ΔH for a transition from one phase to another or for a chemical reaction. They also provide data for the heat capacities C_V at constant volume and C_P at constant pressure. What we measure in a calorimeter at constant volume or in a calorimeter at constant pressure is

$$C_V = \left(\frac{\partial E}{\partial T}\right)_V ; \qquad C_P = \left(\frac{\partial H}{\partial T}\right)_P = \left(\frac{\partial E}{\partial T}\right)_P + P\left(\frac{\partial V}{\partial T}\right)_P . \tag{3.6}$$

In a constant pressure calorimeter we measure indeed the energy increase per degree reduced by the work $-P(\partial V/\partial T)_P$ done by the calorimeter.

In calculations we use as a rule the <u>molal</u> heat capacities obtained from (3.6) through division by the number of moles of the phase. The term specific heat is used for the heat capacity of 1 gram. The total heat capacity of an object, e.g., of a calorimeter, can be obtained by simple addition of the values for its parts, as long as no interfacial interaction, no phase transition, and no heterogeneous reaction occurs. As an illustration we consider a closed vessel with water and ice at 0°: It has no well-defined heat capacity since all energy introduced is used up to melt the ice so that no temperature increase results.

Frequently a heat capactiy is defined as the heat required to increase the temperature by 1°. Again this is not wrong but too narrow because we can increase the energy of an object adiabatically by means of an electric coil.

We need not an arbitrary crutch such as saying that the electrical energy is transformed into heat in the coil.

Differentiation of (3.5) furnishes, in view of (3.4),

$$
\begin{aligned}
C_V &= \left(\frac{\partial E}{\partial T}\right)_V = -T\left(\frac{\partial^2 A}{\partial T^2}\right)_V = T\left(\frac{\partial S}{\partial T}\right)_V ; \\
C_P &= \left(\frac{\partial H}{\partial T}\right)_P = -T\left(\frac{\partial^2 G}{\partial T^2}\right)_P = T\left(\frac{\partial S}{\partial T}\right)_P .
\end{aligned}
\tag{3.7}
$$

A large part of our knowledge in chemical thermodynamics rests on experimental and theoretical information regarding heat capacities and their dependence on the temperature. The theoretical information will be briefly discussed in the next chapter.

For practical applications usually free energies are needed. As a rule, therefore, one has to retrace the trail from the heat capacity to the free energy. More simply, one has to integrate Eq.(3.6) in order to obtain E or H and has to add any isothermal changes. Then one can integrate the differential equations (3.5) and thus obtain A or G. For the second integration, we divide (3.5) by T^2 and obtain

$$\frac{E}{T^2} = \frac{A - T\cdot\partial A/\partial T}{T^2} = -\frac{\partial}{\partial T}\left(\frac{A}{T}\right) ; \quad \frac{H}{T^2} = \frac{G - T\cdot\partial G/\partial T}{T^2} = -\frac{\partial}{\partial T}\left(\frac{G}{T}\right) \tag{3.8}$$

so that with the integration constants A_o and G_o

$$A = -T\int(E/T^2)dT + A_o; \qquad G = -T\int(H/T^2)dT + G_o. \tag{3.9}$$

Integrations in the A - E - C_V system are of course to be carried out at constant volume, those in the G - H - C_P system at constant pressure.

The back-tracing is complete with the integration of (3.9). Since two integrations are carried out in either system, two constants are introduced, in accord with the fact that the two laws defined two integral functions E and S. Both will be discussed in some detail in the next chapter.

3.2 VARIOUS DERIVATIVES

In a procedure similar to that of the last section we obtain immediately from (3.2) and (3.3)

$$\left(\frac{\partial A}{\partial V}\right)_T = -P; \qquad \left(\frac{\partial G}{\partial P}\right)_T = V . \tag{3.10}$$

The relations (3.4), (3.9) and (3.10) are the most important ones between the free energies and other variables. There are a few more that are often required. The mixed second derivatives of A and G can be obtained from (3.4) as well as (3.10):

$$\frac{\partial^2 A}{\partial T \partial V} = -\left(\frac{\partial S}{\partial V}\right)_T = -\left(\frac{\partial P}{\partial T}\right)_V ; \qquad \frac{\partial^2 G}{\partial T \partial P} = -\left(\frac{\partial S}{\partial P}\right)_T = \left(\frac{\partial V}{\partial T}\right)_P . \tag{3.11}$$

Differentiation of (3.5) furnishes

$$\begin{aligned} \left(\frac{\partial E}{\partial V}\right)_T &= \left(\frac{\partial A}{\partial V}\right)_T - T\frac{\partial^2 A}{\partial T \partial V} = -P + T\left(\frac{\partial P}{\partial T}\right)_V ; \\ \left(\frac{\partial H}{\partial P}\right)_T &= \left(\frac{\partial G}{\partial P}\right)_T - T\frac{\partial^2 G}{\partial T \partial P} = V - T\left(\frac{\partial V}{\partial T}\right)_P . \end{aligned} \tag{3.12}$$

From (3.6) and (3.12) we derive

$$\begin{aligned} \left(\frac{\partial C_V}{\partial V}\right)_T &= \frac{\partial^2 E}{\partial T \partial V} = \frac{\partial}{\partial T}\left(\frac{\partial E}{\partial V}\right) = T\left(\frac{\partial^2 P}{\partial T^2}\right)_V ; \\ \left(\frac{\partial C_P}{\partial P}\right)_T &= \frac{\partial^2 H}{\partial T \partial P} = \frac{\partial}{\partial T}\left(\frac{\partial H}{\partial P}\right) = -T\left(\frac{\partial^2 V}{\partial T^2}\right)_P . \end{aligned} \tag{3.13}$$

The difference

$$C_P - C_V = \left(\frac{\partial H}{\partial T}\right)_P - \left(\frac{\partial E}{\partial T}\right)_V = \left(\frac{\partial E}{\partial T}\right)_P - \left(\frac{\partial E}{\partial T}\right)_V + P\left(\frac{\partial V}{\partial T}\right)_P \tag{3.14}$$

is often required. Since E depends only on two variables we can write

$$dE = \left(\frac{\partial E}{\partial T}\right)_V dT + \left(\frac{\partial E}{\partial V}\right)_T dV \tag{3.15}$$

so that the condition of constant P results in

$$\left(\frac{\partial E}{\partial T}\right)_P = \left(\frac{\partial E}{\partial T}\right)_V + \left(\frac{\partial E}{\partial V}\right)_T \cdot \left(\frac{\partial V}{\partial T}\right)_P . \tag{3.16}$$

Introducing (3.12) we have

$$\left(\frac{\partial E}{\partial T}\right)_P = \left(\frac{\partial E}{\partial T}\right)_V + \left[-P + T\left(\frac{\partial P}{\partial T}\right)_V\right]\left(\frac{\partial V}{\partial T}\right)_P \tag{3.17}$$

and (3.14) becomes

$$C_P - C_V = T\left(\frac{\partial P}{\partial T}\right)_V \cdot \left(\frac{\partial V}{\partial T}\right)_P . \tag{3.18}$$

These are only a few of a very large number of relations between the thermodynamic variables. A method of developing them from a fairly small table has been presented by Bridgman [31].

There is no need to memorize the equations discussed in this chapter. The expert will be able to derive them without consulting a book. One should know that they exist and where to look for them. It helps in understanding and applications to know that the temperature dependence of the free energy leads always to a calorimetrically determined quantity (E and C_V or H and C_P), and that the pressure dependence of G is given by V.

All relations derived here for extensive quantities G, H, V and so on apply, of course, immediately to the corresponding molal quantities since they are valid for any amount of the substance. Moreover, since the differentiation indicated in Eq.(1.18) is independent of the differentiations in this section, we can always change the order of differentiation and thus obtain the resulting equations for partial molal quantities $\overline{G}_j$, $\overline{H}_j$ and so on.

3.3 THE EQUATION OF STATE

Since an object under the restrictions of this chapter has only two independent variables, there must exist a relation between the quantities P, V and T. It is called the equation of state. The nature of this relation is individually different for each substance. But in the gaseous state at high dilution specific interactions between molecules are weak because the average distance between them is large. Thus the pressure exerted by the impact of the molecules on the wall of the confining vessel depends only on the hitting molecules, and more specifically on their mass. For this reason a very general and very simple theory leads to the equation of state for the perfect or ideal gas. The problem of the equation of state is an example for both the usefulness and the limitations of molecular theory in practical applications.

3.31 The Perfect Gas. The kinetic theory of gases is based on the tenet that the temperature is a measure of the intensity of molecular motion, more precisely that it is proportional to the average kinetic energy of the molecules. The collisions of the molecules with the enclosing vessel produce the pressure exerted on it.

The model shown in Fig. 3.1, though extremely simplified, demonstrates the essential features: A cube of the volume $\overline{V}(cm^3)$ contains one mole of a gas, or N_L molecules (N_L = Avogadro's constant or Loschmidt's number per mole). The pressure is the force exerted by the molecules on 1 cm^2 of the wall. It is equal to the change of linear momentum with time, i.e., to the momentum change perpendicular to the wall of all molecules arriving there within one second. In order to estimate the number of these molecules we replace the unordered motion by assuming for the model that 1/6 of all molecules travel with a mean velocity v in the + x-direction, another 1/6 in the - x-direction and so on. Therefore 1/6 of the molecules in the small prism of volume $1 \times 1 \times v = v(cm^3)$ hit the 1 cm^2 square every second perpendicularly, each with the linear momentum $(\overline{M}/N_L)\cdot v$ ($\overline{M}$ being the molal weight). There are $v\cdot N_L/V$ molecules in this prism. The molecules are elastically

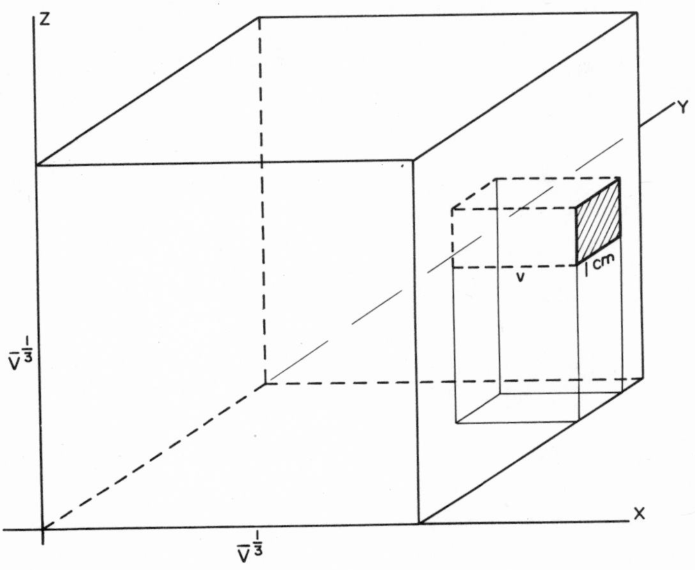

Fig. 3.1. The Perfect Gas

reflected and leave the wall with the same velocity in the opposite direction so that the change of momentum for each molecule is $2\bar{M}v/N_L$. The total change of momentum in a second per cm^2, or the pressure, is therefore

$$P = \frac{vN_L}{6\bar{V}} \cdot \frac{2\bar{M}v}{N_L} = \frac{2}{3\bar{V}} \cdot \frac{\bar{M}v^2}{2} . \tag{3.19}$$

Since the average kinetic energy is assumed proportional to the temperature, we may introduce a proportionality factor R by

$$RT = (2/3) \cdot (\text{average kinetic energy per mole}) \tag{3.20}$$

and obtain

$$P\bar{V} = RT . \tag{3.21}$$

This relation expresses the fact, demonstrated in molecular theory, that the average kinetic energy for one mechanical degree of freedom is the same for all substances. The perfect gas equation is the oldest tool of ascertaining the molal weight of a substance. Since the molal volume $\bar{V}$ for a given pressure and temperature is given by (3.21), the molal weight is the weight of the substance contained in the volume calculated from (3.21) for the measured temperature and pressure.

The perfect gas equation is an excellent example of an extremely useful limiting law. Its accuracy increases with decreasing pressure. Sometimes one hears the objection "But a perfect gas does not exist." True; but the accuracy is amply sufficient in numerous problems and very often even a crude approximation is useful. The appropriate use of such a relation requires some information of its accuracy. Actually the deviations at room temperature and atmospheric pressure amount in general to a few per cent. The following section will show more details.

3.32 Real Gases. For the efficient survey of a large amount of information, particularly by means of diagrams, the selection of a deviation function is often very useful. Such a function is the compressibility factor

$$Z = P\bar{V}/(RT). \tag{3.22}$$

Obviously it is unity for a perfect gas so that its deviation from unity represents the deviation of a gas from the perfect case.

The "generalized chart" in Fig. 3.2 comprises a great deal of information. It shows Z for a number of isotherms as a function of the pressure. The temperature in this diagram is indicated as the reduced temperature $T_r = T/T_c$ or the quotient of the temperature and the critical temperature T_c. The abscissa is the reduced pressure $P_r = P/P_c$. The critical quantities will be discussed a little later.

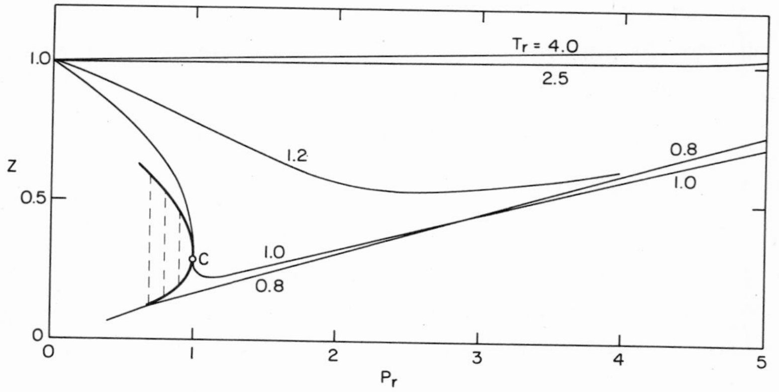

Fig. 3.2. Compressibility Factor Z (data from Pitzer's Tables; the broken lines are liquid vapor tie lines. C critical point)

The diagram shows $Z = 1$ at $P_r = 0$ for all isotherms. The deviations at low pressures can be approximated by the limiting tangents. They can be represented by means of a temperature dependent coefficient β as

$$Z = 1 + \frac{\beta P_c}{RT} \cdot P_r = 1 + \frac{\beta P}{RT} . \tag{3.23}$$

The slope β of the limiting tangents is approximately equal to the so-called second virial coefficient β' in the series

$$P = \frac{RT}{\bar{V}}\left(1 + \frac{\beta'}{\bar{V}} + \frac{\gamma'}{\bar{V}^2} + \cdots\right) . \tag{3.24}$$

This representation, called for weak historical reasons the virial series, has the advantage that one can attribute a molecular-theoretical meaning to β' and even γ'. It is not appropriate for high pressures; actually it contradicts an important fact, namely, that the pressure increases without limit when the volume approaches a finite limit b.

3.33 Two-Parameter Equations. The famous equation of van der Waals

$$P = \frac{RT}{\bar{V}-b} - \frac{a}{\bar{V}^2} \tag{3.25}$$

has the advantages of a simple theoretical basis and of elucidating the whole field of the gaseous and liquid states.

The two reasons for deviations from the perfect gas equation at moderate pressures had been discussed even before van der Waals. The volume available to a molecule for its motion is the total volume $\bar{V}$ reduced by the volume b occupied by all other molecules present in the same volume $\bar{V}$. The true concentration in the freely available space is therefore $N_L/(\bar{V}-b)$ rather than $N_L/\bar{V}$. Therefore the term $RT/\bar{V}$ of the perfect gas is replaced by $RT/(\bar{V}-b)$ in (3.25). The second reason is the intermolecular attraction (which leads to condensation at high pressures); it increases with decreasing distances between the molecules. Its effect is to pull back the molecules from the wall to the bulk. Thus it results in a reduction of the pressure; for each molecule near the wall of the vessel the effect is proportional to the concentration of the retracting molecules or to $1/\bar{V}$. Since the concentration of the molecules hitting the wall during a second is also proportional to the concentration, the total effect on the pressure is proportional to its square, or to $1/\bar{V}^2$.

Van der Waals not only derived these two effects but also discussed the consequences of his equation for the equilibrium between gas and liquid, for the critical state, and for the correspondence in the behavior of different substances.

Qualitatively the equation was tremendously successful. The quantitative representation of experimental data was unsatisfactory. For this reason, applications will be discussed for a modification, proposed by Redlich and Kwong [32],

$$P = \frac{RT}{\bar{V}-b} - \frac{a}{T^{0.5}\bar{V}(\bar{V}+b)} \quad . \tag{3.26}$$

There is no really good theoretical justification for the two changes in the denominator of the a-term but the agreement with observed data is considerably improved so that (3.26) is sufficient for a number of practical applications.

The pressure of n-butane is shown in Fig. 3.3 as a function of the molal

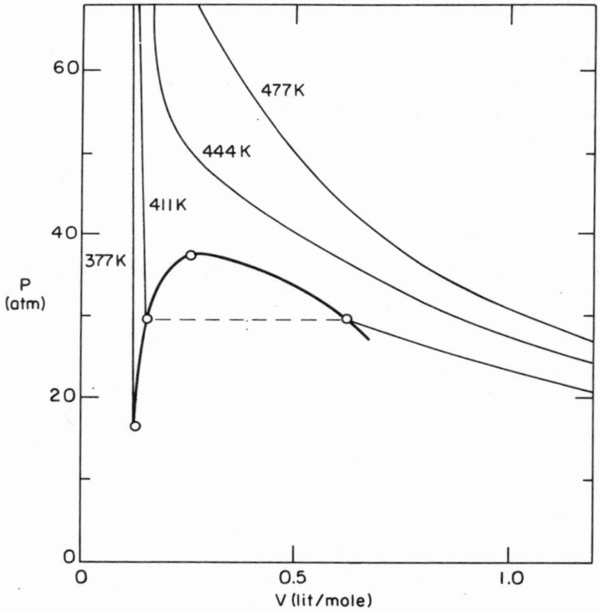

Fig. 3.3. Isotherms and Saturation Line of n-Butane (data from Olds, Reamer, Sage and Lacey, Ind. Eng. Chem., 36 (1944) 282)

volume for several temperatures. Since an equation of the type of (3.25) or (3.26) is of third order in $\bar{V}$, it is satisfied (for prescribed values of P and T) either by a single value of $\bar{V}$ or by three values. Any horizontal line in Fig. 3.3, prescribing a value of P, cuts any isotherm above 425.16 K only once. But horizontal lines below P = 37.47 atm cut isotherms for temperatures lower than 425.16 K three times. In this case the lowest of the three volumes is that of the liquid in equilibrium with its vapor (bubble point); the highest of the three volumes is the volume of the saturated vapor (dew point). The middle volume has no actual meaning.

The diagram expresses the fact that equilibrium between the two phases requires equality of temperature as well as of pressure. The saturation line (bubble point line and dew point line) connects the points of all states that are in equilibrium with another phase. Two phases in equilibrium are indicated by a tieline (horizontal in Fig. 3.3). The region below the saturation line is heterogeneous: A point here means equilibrium between the liquid and the vapor indicated by the end points of the tieline. The relative amounts of the two phases vary along the tieline, but the intensive properties of each of the phases are unchanged since they depend only on temperature and pressure.

The critical point C separates the isotherms representing only gaseous states from those representing also liquids. A critical state is generally defined as the common limit of two phases in equilibrium. The critical values P_c, $\overline{V}_c$, T_c of P, $\overline{V}$, T are characteristic properties of a substance. The diagram shows that the critical isothermal has a horizontal inflection tangent in the critical point, i.e., we have not only

$$\left(\frac{\partial P}{\partial \overline{V}}\right)_c = 0 \tag{3.27}$$

but also [33]

$$\left(\frac{\partial^2 P}{\partial \overline{V}^2}\right)_c = 0. \tag{3.28}$$

These are the conditions for a critical liquid-vapor point.

The critical data represent an important tool for characterizing the behavior of a substance. The equations (3.27) and (3.28) are satisfied by (3.26) if we put

$$a = 0.42748\ R^2T_c^{2.5}/P_c\ ; \tag{3.29}$$

$$b = 0.2599\ \overline{V}_c = 0.08664\ RT_c/P_c. \tag{3.30}$$

The last relation is the strongest and the weakest feature of (3.26). It is a well established experimental fact that the limiting volume b is about 0.26 $\overline{V}_c$. But (3.30) implies

$$Z_c = P_c\overline{V}_c/RT_c = 1/3 \tag{3.31}$$

and this value is a poor approximation (see Table 3.1).

TABLE 3.1

Critical Parameters

Substance	T_c K	P_c atm	Z_c	ω
Oxygen	154.9	50.14	0.294	0.021
Ethane	305.20	48.43	0.282	---
Ethylene	283.06	50.50	0.269	0.111
Carbon tetrachloride	556.3	45.0	0.272	0.205
Benzene	562.61	48.6	0.274	0.210
Carbon dioxide	304.19	72.85	0.276	0.225
Sulfur dioxide	430.4	77.7	0.269	0.270
n-Hexane	507.30	29.71	0.264	0.290
Methanol	512.58	78.50	0.200	0.556

The results expressed in (3.29) and (3.30) enable us to replace a and b in (3.26) by P_c and T_c. It is convenient to introduce the reduced variables

$$P_r = P/P_c; \qquad T_r = T/T_c; \qquad V_r = \bar{V}/\bar{V}_c. \tag{3.32}$$

With (3.29), (3.30) and (3.32), the equation of state (3.26) becomes

$$P_r = \frac{3T_r}{V_r - 0.2599} - \frac{3.8473}{T_r^{0.5}\, V_r(V_r + 0.2599)} . \tag{3.33}$$

For computations, one replaces V_r usually by Z and Z_c as given in (3.22) and (3.31). The result is

$$Z = \frac{Z}{Z - 0.08664\, P_r/T_r} - \frac{0.42748\, P_r}{T_r^{2.5}(Z + 0.08664\, P_r/T_r)} . \tag{3.34}$$

This implicit relation for Z can be solved in a step-by-step approximation procedure. The normal form of the cubic equation for Z is

$$Z^3 - Z^2 + (ZP_r/T_r)(0.42748/T_r^{1.5} - 0.08664 - 0.007506\, P_r/T_r)$$
$$- 0.037037\, P_r^2/T_r^{3.5} = 0. \tag{3.35}$$

One may solve for Z by Cardan's procedure; computer subroutines for the solution are common.

This relation, as well as (3.33), expressed by reduced variables, does not contain any individual parameters (except T_c and P_c which are implicit in the reduced variables). It should therefore be valid for any substance. This is the theorem of corresponding states of van der Waals (it is implied in any two-parameter equation). Practically the theorem furnishes a good approximation for low-molecular, non-polar substances. For water and other highly associated substances, for hydrogen and helium the deviations are large. In general, the deviation of Z_c from the value 0.291 gives a fair idea how far the substance deviates from the theorem of corresponding states (see Table 3.1).

In order to calculate the second virial coefficient we note that according to (3.23) for small values of P

$$\beta = RT(Z-1)/P = \bar{V}(Z-1)/Z. \tag{3.36}$$

Multiplying (3.26) by $\bar{V}/RT$, we obtain

$$Z = \frac{\bar{V}}{\bar{V}-b} - \frac{a}{RT^{1.5}(\bar{V}+b)}$$

$$Z - 1 = \frac{b}{\bar{V}-b} - \frac{a}{RT^{1.5}(\bar{V}+b)} . \tag{3.37}$$

This relation is introduced into (3.36) so that in sufficient approximation for large values of $\bar{V}$ and with the numerical values given by (3.29) and (3.30)

$$\beta = \frac{1}{Z}\left[b \cdot \frac{\bar{V}}{\bar{V}-b} - \frac{a}{RT^{1.5}} \cdot \frac{\bar{V}}{\bar{V}+b}\right] = b - \frac{a}{RT^{1.5}}$$

$$= 0.08664\ (RT_c/P_c)[1 - 4.934(T_c/T)^{1.5}]. \tag{3.38}$$

For nonpolar substances at low temperatures ($T < 0.8\ T_c$) a similar relation proposed by D. Berthelot

$$\beta = 0.0703(RT_c/P_c)[1-6(T_c/T)^2] \tag{3.39}$$

gives better results. Both equations are unsatisfactory for highly polar substances.

The second virial coefficient assumes negative values for moderate and low temperatures, positive values for high temperatures. The temperature for which $\beta = 0$ (horizontal limiting tangent in Fig. 3.2) is called the Boyle temperature. Here the deviations from the perfect gas law are small over a fairly large range (nitrogen at 49°C). From (3.38) we conclude that the Boyle temperature is

$$T_B = 2.90\ T_c , \tag{3.40}$$

a crude approximation for nonpolar gases.

Equation (3.26) represents the bulk of observed data almost as well as the generalized charts, which are the result of a judicious averaging of available experimental data. An equation containing only two individual parameters is necessarily in accord with the theorem of corresponding states as do the generalized charts. Therefore no such equation can do better than the charts.

3.34 Other Equations of State. Generalized charts and other two-parameter equations have long been known to be not generally satisfactory. It has been obvious to attempt an improvement by introducing a third individual parameter. The success of Pitzer [34] and his coworkers shows that reasonably good representation can be obtained if water, hydrogen and helium are excluded.

Pitzer extended the idea of the generalized charts. He and his coworkers found that a satisfactory representation of the compressibility factor

$$Z = Z_0 + \omega Z_1 \tag{3.41}$$

can be achieved by means of two tabulated functions Z_0 and Z_1 of T_r and P_r. The parameter ω is called "acentric factor" because it increases with the deviation of the molecule from spherical shape. It is the third parameter (in addition to T_c and P_c). A similar system has been developed by Riedel [35].

The acentric factor is experimentally defined by means of the vapor pressure $\bar{p}$ of the liquid substance at the reduced temperature 0.7 as

$$\omega = \log (P_c/\bar{p}) - 1.000. \tag{3.42}$$

The critical compressibility factor Z_c does about as well as the third parameter. The connection between the two parameters is

$$Z_c = 0.291 - 0.08\omega \tag{3.43}$$

according to Pitzer's tables [34,36] (cf. Table 3.1).

These tables are very useful for occasional estimates of a single value or a few values of Z. For routine work the double interpolation with respect to T_r and P_r is not convenient. Moreover, an algebraic representation is desirable for the use in calculating fugacity coefficients, especially of the constituents of solutions. The availability of automatic computers also favors algebraic representations.

An equation proposed by Benedict, Webb and Rubin [37] has been widely used. A disadvantage is the implication of a limiting value zero of the volume at high pressure. The high flexibility achieved by eight individual coefficients ensures a good representation of observed data but makes an extrapolation beyond the experimental range risky.

"Expansible" equations have been proposed by Martin and Hou [38] and by Black [39]. They suggested reasonable series with terms of decreasing significance. All available data may be used for the computation of the coefficients. The method is useful for some practical problems.

The extension of Eq.(3.26) by addition of one or two parameters has been proposed by a number of authors. The problem is attractive because Eq.(3.26) offers a good starting point. Actually it turns out to be more difficult than expected so that the best solution is not yet manifest.

The P-V-T relation of liquids and solids is of much lesser interest than that of gases because of the small range of volume variation. An equation proposed by Tait (1888) for liquids

$$V = A + C \log [B(T) + P] \qquad (3.44)$$

has been found useful at times.

3.35 Joule-Thomson Coefficient. If the molecules of a perfect gas are flowing through a hole from one vessel to another, their average velocity does not change since they are under no other influence but elastic collisions. In view of Eq.(3.20) the temperature of a perfect gas does not change in an expansion of this kind, i.e., without introduction of heat from the outside.

In a real gas, however, the molecules must overcome attraction forces acting between them and are helped in the expansion by repulsive forces. Accordingly the temperature on expansion may change. The work required to pass one mole (initial molal volume $\bar{V}_I$) under the constant pressure P_I through the hole is $P_I\bar{V}_I$, but on the other side the outflowing gas does the work $P_F\bar{V}_F$ against the constant final pressure P_F. We keep the gas thermally insulated. The corresponding (adiabatic-irreversible) energy increase of one mole of the gas

$$\Delta\bar{E} = P_I\bar{V}_I - P_F\bar{V}_F = -\Delta(P\bar{V}) . \qquad (3.45)$$

According to the definition of the heat content (2.34) the change in heat content ΔH is therefore zero during adiabatic expansion. The specific temperature change

$$\mu = (dT/dP)_H \qquad (3.46)$$

is called Joule-Thomson coefficient. The state of one mole of a gas is determined by two independent variables; we choose T and P, so that constancy of the heat content is expressed by

$$d\bar{H} = \left(\frac{\partial\bar{H}}{\partial T}\right)_P dT + \left(\frac{\partial\bar{H}}{\partial P}\right)_T dP = 0. \qquad (3.47)$$

Therefore we obtain

$$\mu = -\left(\frac{\partial \overline{H}}{\partial P}\right)_T \Big/ \left(\frac{\partial \overline{H}}{\partial T}\right)_P = -\left[\overline{V} - T\left(\frac{\partial \overline{V}}{\partial T}\right)_P\right] \Big/ C_P \qquad (3.48)$$

according to (3.6) and (3.12).

For a perfect gas we derive easily from Eq.(3.21) that

$$\partial \overline{V}/\partial T)_P = R/P = \overline{V}/T \qquad (3.49)$$

so that $\mu = 0$, as expected. For real gases we can calculate the product

$$\mu \overline{C}_P = T\left(\frac{\partial \overline{V}}{\partial T}\right)_P - \overline{V} \qquad (3.50)$$

from any equation of state. From the definition (3.22) of the compressibility factor we derive

$$\left(\frac{\partial Z}{\partial T}\right)_P = \frac{P}{R}\left[-\frac{\overline{V}}{T^2} + \frac{1}{T}\left(\frac{\partial \overline{V}}{\partial T}\right)_P\right] = \frac{P}{RT^2}\left[T\left(\frac{\partial \overline{V}}{\partial T}\right)_P - \overline{V}\right] \qquad (3.51)$$

so that

$$\mu \overline{C}_P = \frac{RT^2}{P}\left(\frac{\partial Z}{\partial T}\right)_P. \qquad (3.52)$$

At low pressure, where the second virial coefficient β is sufficient to represent Z, we find from (3.23)

$$\left(\frac{\partial Z}{\partial T}\right)_P = \frac{P}{RT^2}\left(T\frac{\partial \beta}{\partial T} - \beta\right) \qquad (3.53)$$

$$\mu \overline{C}_P = T\frac{\partial \beta}{\partial T} - \beta . \qquad (3.54)$$

The Joule-Thomson effect is technically interesting because of its application in refrigeration, which depends primarily on the temperature drop in the irreversible expansion of a gas. Heat capacity and equation of state furnish a complete description of the cooling effect.

Adiabatic expansion does not always result in cooling. Hydrogen, for instance, becomes warmer on expansion at room temperature. The temperature at which the Joule-Thomson coefficient goes through zero is called inversion temperature; according to (3.52) it is characterized by

$$(\partial Z/\partial T)_P = 0. \qquad (3.55)$$

This condition is satisfied in a diagram such as a generalized chart (Fig. 3.2) where vicinal isotherms of Z are crossing. One point of this kind can be seen for $T_r = 0.8$ and $T_r = 1$ at a reduced pressure below 3. Obviously Z does not change with T at constant P in such a point. These inversion points form a curve (inversion curve) in the diagram.

The limit of the inversion temperature at low pressure is sometimes called Joule-Thomson temperature. We can estimate this temperature by means of an equation of state such as (3.26). Differentiating (3.38) with respect to T and using (3.54), we obtain

$$\mu C_p \;=\; 0.08664 \;\; \frac{RT_c}{P_c}\left[12.335\left(\frac{T_c}{T}\right)^{1.5} - 1\right] . \qquad (3.56)$$

The inversion temperature at low pressure is therefore

$$T_J \;=\; 5.334\; T_c \; . \qquad (3.57)$$

The observed data, according to a recent survey [40] are shown in Table 3.2. Except for helium and hydrogen, the relation (3.57) furnishes a fair estimate for nonpolar substances. The deviations are considerable for polar substances such as carbon dioxide.

Table 3.2

Inversion at Low Pressure

Substance	T_J(K)	T_J/T_C
Helium	46.3	8.87
Neon	228	5.13
Argon	768	5.09
Krypton	1079	5.15
Xenon	1476	5.09
Hydrogen	204	6.14
Deuterium	209	5.46
Nitrogen	607	4.81
Carbon monoxide	644	4.85
Carbon dioxide	1275	4.19
Methane	953	5.00

3.4 THE FUGACITY COEFFICIENT

It is the dependence of the free energy on the pressure that justifies the extraordinary effort spent by numerous authors on the problem of the equation of state; the relation between pressure, volume and temperature in itself would not be interesting enough. But according to (3.10) we need the knowledge of the volume as a function of pressure at the temperature under consideration to find the molal Gibbs free energy

$$\overline{G} \;=\; \overline{G}^o + \int_{P^o}^{P} \overline{V} dP \qquad (3.58)$$

as a function of the pressure. The superscript ^{o} indicates the standard state, i.e., P^o is the standard pressure at which $\overline{G}$ has the standard or reference value $\overline{G}^o$. Usually P^o is chosen as 1 atm. The quantity $\overline{G}^o$ is an integration constant with respect to P; but it varies with the temperature. As a temperature function, $\bar{G}^o$ is computed with the aid of Eq.(3.9) or an equivalent relation. Eq.(3.58) furnishes the pressure dependence for any temperature.

For the convenience of thermodynamic calculations Lewis introduced the fugacity f, which will be replaced here for gaseous substances with the aid of the fugacity coefficient

$$\phi = f/P. \tag{3.59}$$

The definition of the fugacity coefficient (and the fugacity) is given by

$$\bar{G} = \bar{G}^o + RT\,\ell n(P\phi). \tag{3.60}$$

Differentiation furnishes

$$\partial\bar{G}/\partial P = RT\,\partial\ell n(P\phi)/\partial P = \bar{V}. \tag{3.61}$$

Replacing the volume by the compressibility factor (3.22) we obtain

$$\frac{\partial\ell n\phi}{\partial P} = \frac{\bar{V}}{RT} - \frac{1}{P} = \frac{Z-1}{P} \tag{3.62}$$

so that the fugacity coefficient can be derived from any equation of state by

$$\ell n\phi = \int_{P^o}^{P} \frac{Z-1}{P}\,dP = \int_{P_r^o}^{P_r} \frac{Z-1}{P_r}\,dP_r. \tag{3.63}$$

In the range of moderate pressures, i.e., as long as the second virial coefficient furnishes a sufficient approximation of Z, we may write, using (3.23),

$$\ell n\phi = \frac{\beta P_c}{RT} \cdot (P_r - P_r^o) = \frac{\beta}{RT}(P - P^o). \tag{3.64}$$

Often P^o will be negligible compared with P. For calculations of high precision we abandon the convention $P^o = 1$ and choose P^o so low that $\beta P^o/RT$ is negligible. Thus we have for moderate pressures always

$$\ell n\phi = \beta P/(RT) \tag{3.65}$$

and the fugacity coefficient at the standard state is unity. In other words, the fugacity of a gas in the standard state is equal to the pressure.

Each equation of state furnishes a function for the fugacity coefficient. From (3.26) one derives

$$\ell n\phi = Z-1 - \ell n(Z - 0.08664\,P_r/T_r) - (4.934/T_r^{1.5})\ell n[1 + 0.08664\,P_r/(T_r Z)]. \tag{3.66}$$

For a verification we differentiate this equation with respect to P_r and obtain an equation of the form

$$(\partial\ell n\phi/\partial P_r)_{T_r} = J\cdot\partial Z/\partial P_r + L. \tag{3.67}$$

Then we see from (3.34) that J = 0, and that

$$L = (Z - 1)/P_r \tag{3.68}$$

as required by (3.63).

A generalized diagram of fugacity coefficients is shown in Fig. 3.4. Fugacity coefficients can be derived in a similar way from any equation of state. But the calculation difficulties increase of course when the equation of state becomes more complicated.

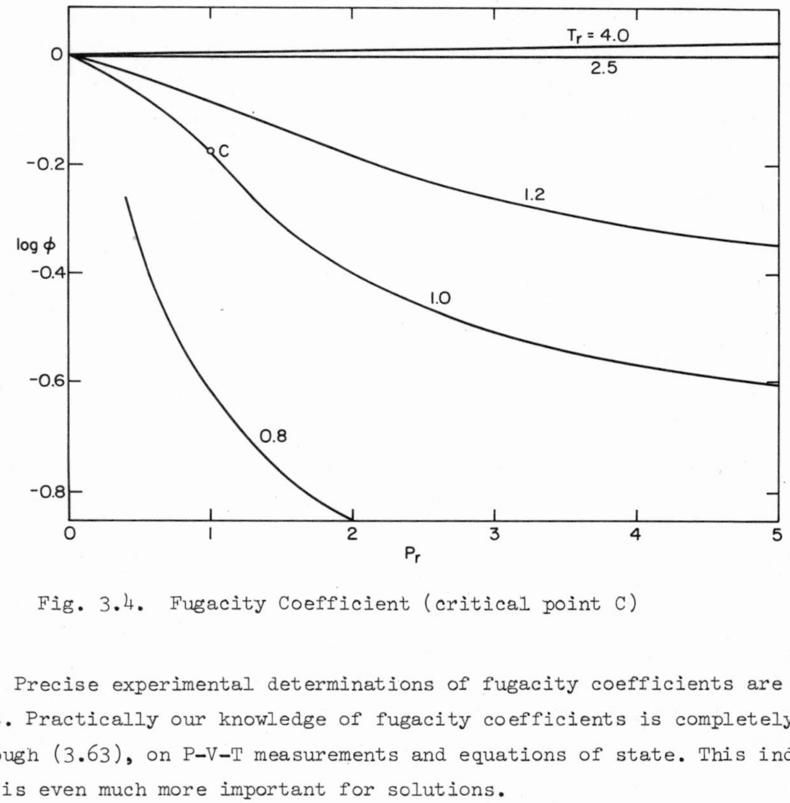

Fig. 3.4. Fugacity Coefficient (critical point C)

Precise experimental determinations of fugacity coefficients are difficult. Practically our knowledge of fugacity coefficients is completely based, through (3.63), on P-V-T measurements and equations of state. This indirect way is even much more important for solutions.

SUMMARY OF THE THIRD CHAPTER

The temperature dependence of a free energy can be represented by the heat content or the entropy, further by a heat capacity. The change of the Gibbs free energy with pressure is given by the volume. These and a number of other relations are important tools for the consistent tabulation of thermodynamic properties as functions of the temperature and pressure.

The equation of state is a necessary relation between temperature, pressure and molal volume of any substance of a given composition. The equation of the perfect gas is a useful approximation for every molecular species in the state of a dilute gas.

The data and interpretation of the P-V-T relationship are the basis for representing the Gibbs free energy as a function of the pressure. The fugacity coefficient is a useful measure of the deviation of the free energy from that of a perfect gas.

REFERENCES

(31) P. W. Bridgman, Phys. Rev., 3 (1914) 273. A selection can be found in Lewis-Randall-Pitzer-Brewer, Thermodynamics, McGraw-Hill, New York, 1961, Appendix 6.

(32) O. Redlich and J. N.S. Kwong, Chem. Revs., 44 (1949) 233.

(33) This is the traditional second condition for the critical point. Actual experimental data indicate a discontinuity ($\partial^2P/\partial V^2$ jumping from a positive to a negative value) rather than a continuous function through zero.

(34) K. S. Pitzer and co-workers. See Lewis-Randall-Pitzer-Brewer, Appendix I.

(35) L. Riedel, Chem. Ing. Tech., 26 (1954) 83, 259, 679; 27 (1955) 209, 475; 28 (1956) 557.

(36) J. M. Prausnitz, Molecular Thermodynamics of Fluid-Phase Equilibria, Prentice-Hall, Englewood, N. J., 1969; p. 129.

(37) M. Benedict, G. B. Webb and L. C. Rubin, J. Chem. Phys., 8 (1940) 334; 10 (1942) 747.

(38) J. J. Martin and Y-C. Hou, AIChE Journ., 1 (1955) 142.

(39) C. Black, Ind. Eng. Chem., 50 (1958) 391.

(40) D. Straub, A. Schaber and T. E. Morsy, Kältetechnik, 17 (1965) 212.

CHAPTER 4. INFORMATION FROM MOLECULAR THEORY

A coherent structure of thermodynamics can be developed, as the preceding chapters show, without recourse to the statistical branch. But molecular theory and statistical methods lead to models that often depict real substances closely enough to contribute greatly to our understanding. The topics discussed in this chapter will also show the considerable quantitative information that can be derived from theoretical models.

Molecular theory was introduced in a brief excursion as the basis of the perfect gas law. It is surprising how such an important quantitative relation can be deduced from very simple and plausible assumptions. A similar objective in this chapter will be the deduction of the properties of a perfect solution and of the behavior of various substances at low temperatures. Our interest will be concentrated on these concrete problems rather than on a general survey of statistical methods.

The theory is based on an idea proposed by Demokritos twentyfour centuries ago. The properties of an object, as observed and measured by our coarse senses, are the combined effect of an enormous number of molecules. The problem therefore is the deduction of the observable macro-properties from the micro-image of the assembly of molecules.

Maxwell and Boltzmann started from the mechanics of very simple molecular models, depicting the molecules as tiny billiard balls. Gibbs developed statistical mechanics from the general theorems of analytical mechanics. The modern presentation of statistical thermodynamics goes back to Planck in many details. Several books give a good survey of modern statistical thermodynamics [41].

4.1 OBSERVABLE AND MOLECULAR PROPERTIES

More than a century ago, Loschmidt noticed that the number of molecules in one mole can be derived from the kinetic theory of the viscosity of gases. The viscosity of spherical molecules is proportional to their cross section. The proper volume of closely packed molecules can be estimated from the limiting volume of a liquid at low temperature. Viscosity and limiting volume furnish therefore the radius of the molecule, and also the number N_L of the molecules in one mole, which has been called Loschmidt's number per mole or Avogadro's constant. The determination of N_L by a score of widely different methods with a concordant result ($N_L = 6.0222 \times 10^{23}$ mole^{-1}) has been one of the achievements that gave substance to the deductions of molecular theory.

Energy is a property that can be attributed to molecules in the same way as to large objects. If there is no interaction (i.e., in the perfect gas) the energy of a molecule is the same as in the state of isolation, and the energy of the gas is the sum of the energies of the molecules. Except for the perfect gas, there is always attraction and repulsion between molecules; in addition to the energies of the single molecules, the total energy comprises also the energy due to forces acting between the molecules.

The interpretation of entropy is more sophisticated. Entropy was introduced as a property that increases in an irreversible change of a thermally insulated object. For illustration we consider in Fig. 4.1 the expansion of

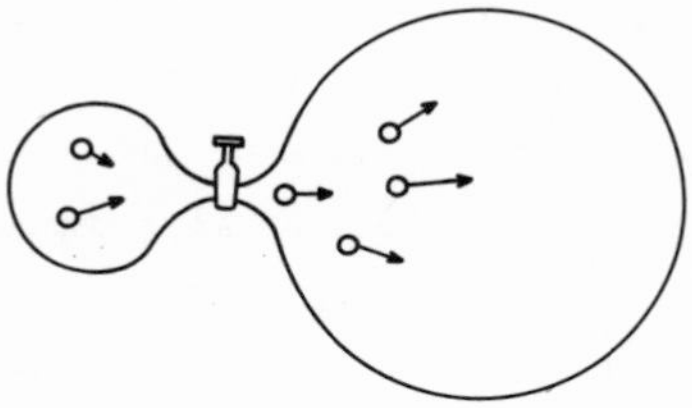

Fig. 4.1. Expansion of a Gas (from the initial volume V_1 to the final total volume V_2 of both vessels)

a mole of a perfect gas from a flask (volume $\bar{V}_1$) to an evacuated flask (final total volume $\bar{V}_2$). By observation we know that this is an irreversible process so that the entropy of the gas increases. What happens if we keep N_L small, rapidly moving billiard balls in the volume $\bar{V}_1$ and then open the stopcock? They stream out, of course, through the stopcock until they are uniformly distributed in the whole vessel. They are not dragged out from $\bar{V}_1$ by a mysterious force exerted by the vacuum. They go into all parts of the vessel because the collisions with other molecules (and the walls) send them forth in all directions. For any single molecule there is a certain probability that it will stay in $\bar{V}_1$ or be sent back to $\bar{V}_1$. In a perfect gas there are no forces (except random forces during collisions) acting on a molecule;

thus one place in the container is just as good for the molecule as any other. The probability for a molecule to be in $\bar{V}_1$ rather than anywhere in the total volume $\bar{V}_2$ is $\bar{V}_1/\bar{V}_2$; for a second molecule it is again $\bar{V}_1/\bar{V}_2$, for both at the same time it is therefore $(\bar{V}_1/\bar{V}_2)^2$, and for all N_L molecules it is $(\bar{V}_1/\bar{V}_2)^{N_L}$. Because N_L is such a large number the probability for all molecules to assemble spontaneously in $\bar{V}_1$ is extremely small as soon as $\bar{V}_1$ is smaller than $\bar{V}_2$ even by as little as one per cent. The relation for the probability ratio

$$W_1/W_2 = (\bar{V}_1/\bar{V}_2)^N \quad (4.1)$$

is illustrated in Table 4.1 for $\bar{V}_1/\bar{V}_2 = 0.99$ and 0.5 and for various values of N. Even for 1000 molecules the probability of assembling in the smaller volume is very small; much more so when N increases to the order of magnitude of $N_L = 6.10^{23}$.

Table 4.1
Probability Ratio for an Expanded Gas

	N	2	3	5	10	100	1000	10 000
$\bar{V}_1/\bar{V}_2=0.99$								
W_1/W_2		0.98	0.97	0.95	0.91	0.37	0.00005	10^{-43}
$\bar{V}_1/\bar{V}_2=0.5$								
W_1/W_2		0.25	0.125	0.032	0.001	10^{-30}	10^{-300}	10^{-3000}

This example illustrates the idea of the probability of a certain state of a molecule and the progress from this to the probability of an assembly, say of one mole. It illustrates an irreversible change as a progress from a state of lower to higher probability.

If we wish to generalize what this example tells us, we should say more definitely how to determine the probability of an assembly. Instead, we shall be satisfied either to introduce ratios of two probabilities by plausible reasoning as we did here, or to call probability the number of different molecular states ("microstates") that are all observed as the same "macrostate" by our senses and measuring instruments.

Clearly we can generalize the example of the expanding gas and say right away that an isolated object spontaneously changes from a state of lower probability to a more probable state. We compare this result with the statement that the entropy of an isolated object increases in an irreversible process. The conclusion is obvious that probability and entropy are essentially identical, i.e., that one is a universal, monotonically increasing function of the other.

The nature of this functional relationship can be easily found. We consider a system consisting of two non-interacting objects. The entropy S of the system is equal to the sum of the entropies S_1 and S_2 of the objects

$$S = S_1 + S_2. \tag{4.2}$$

This follows from the definition of the entropy as the quotient of reversibly introduced heat and temperature. But the probability W of having one object in the state of probability W_1 and another of probability W_2 is (see Appendix 12.1)

$$W = W_1 \cdot W_2. \tag{4.3}$$

The former conclusion that S is a universal function of W can now be expressed by

$$S(W) = S(W_1 \cdot W_2) = S(W_1) + S(W_2). \tag{4.4}$$

The relationship singled out by this condition is the logarithm

$$\ell n(W_1 \cdot W_2) = \ell n\, W_1 + \ell n\, W_2 \tag{4.5}$$

and it can be shown that no other relationship satisfies this condition (Appendix 12.2). Thus we conclude that the entropy is a linear function of ℓn W

$$S = k \ell n\, W + S_o. \tag{4.6}$$

This is Boltzmann's theorem and k is called Boltzmann's constant. The value of k is easily derived from the example of the expanding perfect gas. From (4.6) and (4.1) we find for one mole

$$\bar{S}_1 - \bar{S}_2 = k \ell n(W_1/W_2) = k N_L \ell n(\bar{V}_1/\bar{V}_2). \tag{4.7}$$

We see also from (3.11) that

$$(\partial \bar{S}/\partial \bar{V})_T = (\partial P/\partial T)_V. \tag{4.8}$$

For a perfect gas we have

$$(\partial P/\partial T)_V = R/\bar{V} \tag{4.9}$$

so that

$$\bar{S} = \int \left(\frac{\partial \bar{S}}{\partial \bar{V}}\right)_T d\bar{V} = R \int \frac{d\bar{V}}{\bar{V}} = R \ell n \bar{V} + \text{const.} \tag{4.10}$$

$$\bar{S}_1 - \bar{S}_2 = R \ell n(\bar{V}_1/\bar{V}_2). \tag{4.11}$$

The comparison of this relation with (4.7) gives

$$k = R/N_L. \tag{4.12}$$

Boltzmann's constant is therefore the gas constant per molecule.

The relation between entropy and probability has far reaching consequences in applied thermodynamics.

4.2 THE PERFECT SOLUTION

The idea of the perfect solution, similar to that of the perfect gas, presents a simple model that describes very closely the properties of some mixtures; more often the description is only approximate, but still useful as a starting point for a more accurate representation.

We call a mixture of two substances (Fig. 4.2) A and B perfect if the

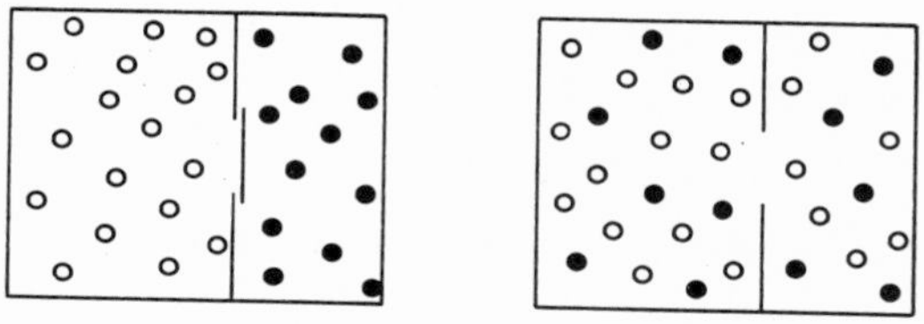

Fig. 4.2. Perfect Solution

forces between two molecules A and A, or B and B, or A and B are equal, and if the sizes of A and B are equal. A perfect solution is very closely realized by a mixture of isotopic molecules: Forces between a benzene molecule and a deuterobenzene molecule are very nearly equal to those between two benzene molecules, and the volumes are also very nearly equal. Deviations from the perfect model are small for mixtures of two similar substances, such as benzene and toluene, o-xylene and m-xylene, butane and isobutane, and so on. They are large for such pairs as hexane-acetone, carbon tetrachloride-methanol, and so on. A mixture of perfect gases is automatically a perfect solution since interacting forces and proper volumes of both components are negligibly small.

In preparing a solution we take molecules A from a pure A-environment and transfer them into a mixed environment of A and B. In a perfect solution the forces acting on the transferred molecule are the same as before. It follows that no energy is liberated or absorbed by making a perfect solution of any concentration from the components. Similarly, it follows from the second requirement that there is no change of the volume. Therefore, there is no change in the heat content H: the heat of mixing is zero.

But the entropy and probability change considerably; mixing, after all, is an irreversible process. We shall estimate the probability change first for a perfect solid solution (perfect mixed crystal). We divide the molal volume of a crystal into N_L equal cubicles and take N_L different molecules that satisfy the requirements of a perfect solution. (In a "thought experiment" we do not worry about trivial difficulties such as finding a very large number of qualified molecules.) Everyone fits just into a cubicle and, because of the assumed equality of intermolecular forces, no molecule prefers any cubicle to any other.

The probability of the state is now given by the number of arrangements which in our measurements appear as the same state of the solution. The first molecule we can place in any of N_L cubicles; for the second we have then $(N_L - 1)$ cubicles open however we have placed the first one. The number of arrangements for two molecules are therefore

$$N_L \cdot (N_L - 1).$$

Obviously we obtain the number of possible arrangements of three molecules by multiplying by $(N_L - 2)$, and so on. The number of arrangements for one mole of the crystal is therefore

$$1 \cdot 2 \cdot 3 \quad \ldots \quad (N_L - 3)\cdot(N_L - 2)\cdot(N_L - 1)\cdot N_L \;=\; N_L! \tag{4.13}$$

(read N_L factorial).

This goes for N_L molecules each of which can be distinguished from all others. If we exchange two molecules, say the first and the second, we have, of course, two different arrangements. But if we now assume that the two molecules are identical, we have only one arrangement instead of the former two. The same holds for any pair of exchangeable arrangements so that the total number with two identical molecules is $N_L!/2$. If three molecules are identical, we retain a single distinguishable arrangement only out of $1\cdot2\cdot3$ original arrangements, and if N_A molecules are identical the total number remaining is

$$N_L!/N_A! \;.$$

(If all molecules are identical, i.e., for a pure substance, the probability is unity because there is only a single arrangement of the N_L molecules.) If in addition N_B other molecules are identical, only 1 out $N_B!$ arrangements is different from the others, and the number of possible arrangements, representing a single observable macro-state is

$$W = \frac{N_L!}{N_A!N_B!} . \tag{4.14}$$

For the mixing of N_A molecules A and N_B molecules B

$$N_A A + N_B B = N_L \quad \text{(Solution AB)}; \quad N_A + N_B = N_L$$

we have therefore

$$\Delta \ln W = \ln W = \ln(N_L!) - \ln(N_A!) - \ln(N_B!). \tag{4.15}$$

In the calculation of $\Delta \ln W$ there is nothing to deduct for the pure substances A and B since the probabilities of the pure substances are unity and their logarithms therefore zero.

Stirling's approximation (discussed in Appendix 12.3) gives for any very large number N the relation

$$\ln(N!) = N\ln N - N. \tag{4.16}$$

Therefore we obtain for the mixing process

$$\Delta W = N_L \ln N_L - N_A \ln N_A - N_B \ln N_B. \tag{4.17}$$

The entropy change ΔS on mixing

$$x_A = N_A/N_L \qquad \text{and} \qquad x_B = 1 - x_B = N_B/N_L \tag{4.18}$$

moles to give one mole of a perfect solution is now given by Boltzmann's relation (4.6) as

$$\begin{aligned} \Delta\bar{S}^* &= k\Delta \ln W = kN_L(\ln N_L - x_A \ln N_A - x_B \ln N_B) \\ &= -R[x_A \ln(N_A/N_L) + x_B \ln(N_B/N_L)] \\ &= -R[x_A \ln x_A + x_B \ln x_B]. \end{aligned} \tag{4.19}$$

The entropy of mixing $\Delta\bar{S}^*$ is necessarily positive since $\ln x_A$ and $\ln x_B$ are always negative. The function $\Delta\bar{S}^*$ is represented in Fig. 4.3.

The core of the argument leading to this important relation is the exchange of molecules leading to different <u>micro</u>states for the same <u>macro</u>-state. Counting the possible exchanges leads to the probability and therefore to the entropy. This count does not depend on our original assumption that the solution is solid. We can count exchanges at a given moment in a liquid or gas just as well, and the result is the same. The relation (4.19) for

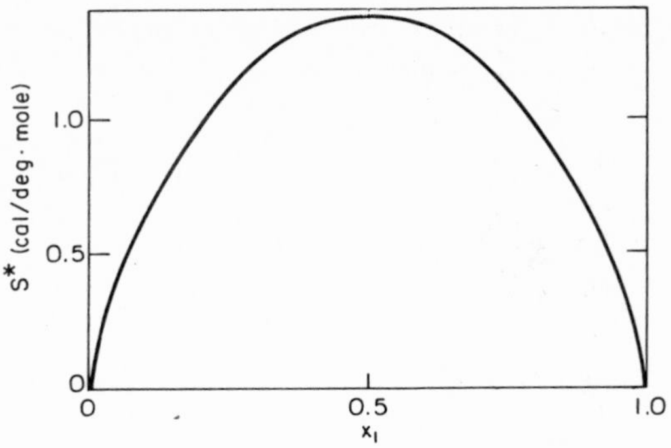

Fig. 4.3. Molal Entropy of Mixing $\bar{S}$* of a Perfect Solution

the entropy of mixing of a perfect solution applies therefore also to liquid and gaseous solutions.

The relation (4.19) contributes always a major part to the entropy change on mixing. Other terms are to be added for deviations from the original model regarding equality of interaction forces and of the volumes. Large deviations may be due to chemical changes linked to the mixing, such as the formation of compounds or dissociation of one component (electrolytic or otherwise). The model of the perfect solution gives us a good standard for comparisons.

The inspection of (4.19) leads to a strange comment, the so-called paradox of Gibbs. The mixing entropy of two similar substances is finite and has the same value however different or similar the two components are. If A is C_6H_6 and if B is C_6D_6 the mixing entropy is given by (4.19). Also if B is C_6HD_5, $C_6H_2D_4$, $C_6H_3D_3$, $C_6H_4D_2$, C_6H_5D the mixing entropy is always the same. But if B is C_6H_6 then the mixing entropy is, of course, zero. There is no continuous transition from the expression in (4.19) to zero.

In fact, the mixing entropy depends on which molecules we call different. If we separate the isotopes of uranium

$$U\ (\text{natural}) = 0.993\ U^{238} + 0.007\ U^{235}$$

the entropy of separation is $-\Delta\bar{S}^*$ as given in (4.19). But as long as the isotopes are not separated they do not contribute to the entropy change of a chemical reaction. For the reaction

$$Na + \tfrac{1}{2}Cl_2 = NaCl$$

the entropy of the mixing of Cl^{35} and Cl^{37} appears in the same way for Cl_2 and NaCl; therefore it cancels in the result. Thus the entropy of mixing can be disregarded as long as no mixing or separation is involved in a reaction. But it has to be taken into account whenever the mixing ratio is changed, however similar the components may be.

The results for a binary solution can be immediately generalized for z components with the numbers of molecules

$$N_1 + N_2 + \dots + N_z = N_L. \tag{4.20}$$

For each additional component j the number of microstates is reduced by a corresponding factor N_j in the denominator of (4.14). The probability of one mole of the solution is therefore given by the relation

$$\Delta \ell n W = N_L \ell n\, N_L - \sum_1^z N_j \ell n\, N_j \tag{4.21}$$

with the condition (4.20), and the entropy of mixing by

$$\Delta \bar{S}^* = -R \sum x_j \ell n\, x_j. \tag{4.22}$$

4.3 THE DISTRIBUTION OF ENERGY

The discussion of the perfect solution illuminated the influence of probability so clearly because in such a solution each arrangement of the molecules is just as good as any other; therefore the probability was found by counting the various arrangements.

A simple count is not sufficient if the exchange of two molecules does not lead to an arrangement of equal likelihood. If helium escapes into the air we may ask for the entropy of the mixture in a tall column. But now the exchange of a helium atom (weight 4) near the bottom for a nitrogen molecule (weight 28) near the top of the column leads to a preferred arrangement. We have to attribute different likelihoods to the two cases. Therefore we have to examine the distribution of molecules between states of different energies.

Planck's quantum theory has shown that there are discrete energy states, i.e., cases in which the energy difference between different states is finite. A continuous sequence of energies exists for the kinetic energy of translation and practically for various limiting cases. It is simpler and for our purpose sufficient to consider a model that provides only discrete energy states. Thus we call j the order number of the state, E_j the energy of a molecule in the state j, N_j the number of molecules in j, and we describe the distribution by the scheme

order number	0	1	2	...	j	...
energy	ε_0	ε_1	ε_2	...	ε_j	...
population number	N_0	N_1	N_2	...	N_j	...
multiplicity	$g_0 = 1$	g_1	g_2	...	g_j	...

The multiplicity or the statistical weight g_j is provided for the case that two or more distinct states happen to have approximately or even exactly the same energy ε_j. The distinction between such "degenerate" states has a theoretical basis. It is seen in certain spectra such as Zeeman spectra, obtained in a magnetic field, which changes the energies of the various degenerate states in different manners and therefore splits a spectral line into a doublet, triplet and so on. In general the assumption $g_0=1$, noted in the array, is permissible. There are a few exceptional cases where other values for g_0 are introduced.

The probability of a distribution of N_L molecules over the energy classes in the preceding schedule presents precisely the same problem as the probability of a mixture containing N_0, N_1, N_2 ... molecules, although the physical meaning is entirely different. The result is

$$\ell n\ W = N_L \ell n\ N_L - \sum g_j N_j \ell n\ N_j . \tag{4.23}$$

The expressions (4.21) and (4.23) are formally different only in the weights g_j because we have to count every energy class j as often as indicated by g_j.

But there is a much more essential difference between the two cases. For a mixture, the numbers of molecules in every class were fixed and given. In the present distribution problem the N_j's are not preset; on the contrary, any molecule can appear in any class j. Now we wish to learn the values of the N_j's when the total energy

$$E = \sum g_j \varepsilon_j N_j \tag{4.24}$$

is prescribed. There is indeed a most likely distribution dependent on the total energy. We have to observe an additional condition, namely, that there is a certain total number of molecules. We consider one mole so that the total number is

$$N_L = \sum g_j N_j . \tag{4.25}$$

It can be shown that deviations from the most likely distribution are observable only in a few exceptional cases, most notably near a critical point.

We derive the most likely distribution by the requirement that W has a maximum value with respect to any set of changes δN_1, δN_2, ... , which satisfies the conditions (4.24) and (4.25). The calculation, which presents no difficulty, is shown in Appendix 11.4.

The result is another law of Boltzmann, namely,

$$N_j = N_0 e^{-(\varepsilon_j - \varepsilon_0)/kT} = N_0 e^{-u_j} \tag{4.26}$$

with the abbreviation

$$u_j = (\varepsilon_j - \varepsilon_0)/kT. \tag{4.27}$$

The condition (4.25) appears now as

$$N_L = N_0 \sum g_j e^{-u_j} = N_0(1 + g_1 e^{-u_1} + g_2 e^{-u_2} + \dots) = N_0 Q. \tag{4.28}$$

The <u>partition function</u>

$$Q = N_L/N_0 = 1 + g_1 e^{-u_1} + g_2 e^{-u_2} + \dots \tag{4.29}$$

plays an important role in the linking of molecular and thermodynamic functions.

In the following the temperature dependence of Q and the thermodynamic functions will be discussed. The volume and all other possible coordinates will be kept constant. Under these conditions the energy ε_j of any state is assumed to be constant. Therefore we have

$$du_j/dT = -(\varepsilon_j - \varepsilon_0)/kT^2 = -u_j/T. \tag{4.30}$$

We derive from (4.29) and (4.26)

$$\begin{aligned} dQ/dT &= g_1 u_1 e^{-u_1}/T + g_2 u_2 e^{-u_2}/T + \dots \\ &= \frac{1}{kT^2} \cdot \sum g_j(\varepsilon_j - \varepsilon_0) e^{-u_j} \\ &= \frac{1}{N_0 kT^2} \cdot \sum g_j(\varepsilon_j - \varepsilon_0) N_j. \end{aligned} \tag{4.31}$$

Writing

$$\bar{E}_0 = \varepsilon_0 N_L \tag{4.32}$$

for the energy of the ground state, we obtain from (4.24)

$$\bar{E} - \bar{E}_0 = \sum g_j \varepsilon_j N_j - \varepsilon_0 \sum g_j N_j = \sum g_j(\varepsilon_j - \varepsilon_0) N_j. \tag{4.33}$$

This result is introduced in (4.31) and furnishes, in view of (4.28),

$$\bar{E} - \bar{E}_0 = N_0 kT^2 \frac{dQ}{dT} = \frac{N_L kT^2}{Q} \cdot \frac{dQ}{dT} = RT^2 \frac{d\ell n\, Q}{dT}. \tag{4.34}$$

For a comparison with (3.8) we write this equation

$$\frac{\bar{E} - \bar{E}_o}{T^2} = -\frac{d}{dT} \frac{\bar{A} - \bar{A}_o}{T}. \tag{4.35}$$

The comparison furnishes

$$\bar{A} - \bar{A}_o = -RT\ell n\ Q = R\ell n(N_O/N_L). \tag{4.36}$$

The partition function Q has therefore a direct thermodynamic significance.

The entropy is now easily identified by means of (3.4) and (4.36) as

$$\bar{S} - \bar{S}_o = -d(\bar{A} - \bar{A}^o)/dT = Rd(T\ell n\ Q)/dT. \tag{4.37}$$

The same result can be obtained by introducing (4.26) into (4.23) and applying Boltzmann's theorem (4.6).

Relation (4.37)

$$\bar{S} - \bar{S}_o = R\ell n\ Q + RT\frac{d\ell n\ Q}{dT} = \frac{\bar{A} - \bar{A}_o}{T} + \frac{\bar{E} - \bar{E}_o}{T} \tag{4.38}$$

leads through (4.36) and (4.35) immediately back to the definition of the free energy A of Helmholtz.

The distribution of the molecules between the various energy classes is illustrated by Figs. 4.4 and 4.5. They show Boltzmann's theorem (4.26)

$$\ell n(N_j/N_O) = -(\varepsilon_j - \varepsilon_o)/kT = -N_L(\varepsilon_j - \varepsilon_c)/RT \tag{4.39}$$

with the molal energy of state j represented by

$$\bar{E}_j - \bar{E}_o = N_L(\varepsilon_j - \varepsilon_o). \tag{4.40}$$

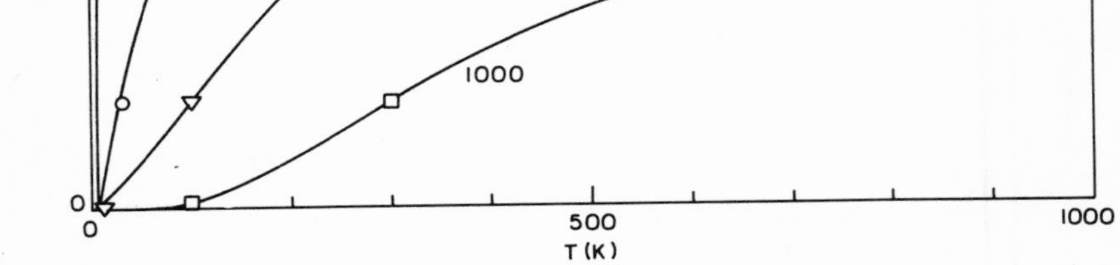

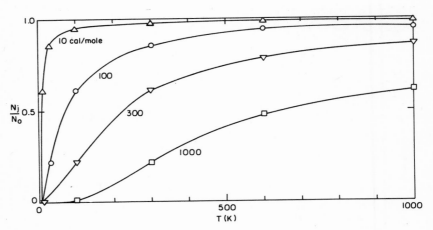

Fig. 4.4. Population of Various Energy Classes (the energy per mole is indicated)

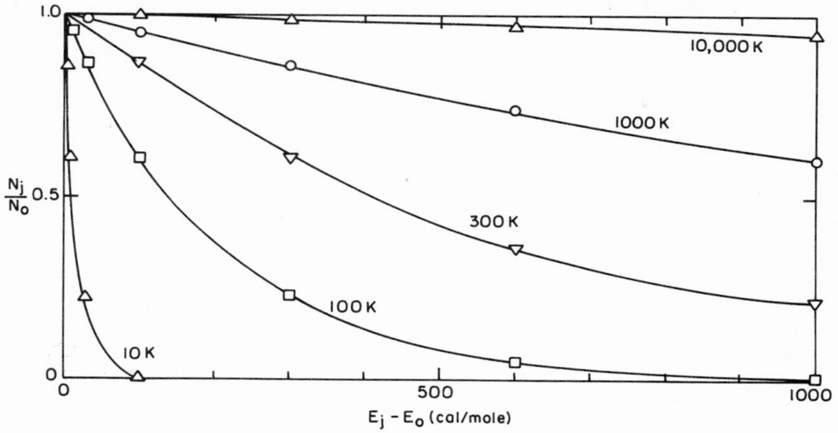

Fig. 4.5. Isotherms for the Population as a Function of the Energy

The ordinate shows the ratio N_j/N_0. It is unity for $\bar{E}_j = \bar{E}_0$, and zero in the limit of T = 0 as can be seen immediately from (4.39). For high temperatures N_j/N_0 tends to unity, rapidly for small values of $\bar{E}_j - \bar{E}_0$, slowly for large ones (Fig. 4.4); in the limit of high temperatures, therefore, all population numbers approach the same value, namely, N_0. At room temperature (around 300 K) states with energies below 300 cal/mole are more or less close to the value of N_0; states above 1000 cal/mole have a low population.

In a similar manner Fig. 4.5 shows that at 10 K only very low states (below 100 cal/mole) are occupied. But at 10 000 K even the state with 1000 cal/mole is close to the limiting value N_0.

The distribution of the energy (Fig. 4.6) looks quite different from the distribution of the population. Taking $(\bar{E}_j - \bar{E}_0)\, N_j/N_0$ as a measure of the energy contained in class j we see that all curves have maxima. For each temperature there exists therefore an energy class which contributes more to the total energy than any other class; from (4.39) we derive easily that this value is

$$\bar{E}_j^* - \bar{E}_0 = RT^* . \tag{4.41}$$

It is actually around the temperature T* that an energy change $\bar{E}_j^* - \bar{E}_0$ produces the most conspicuous effects on some thermodynamic properties.

As a rule, the energy of a molecule consists of various contributions, for instance of one or more rotational energies and one or more vibrational energies. There may be some interaction between the two kinds of energy;

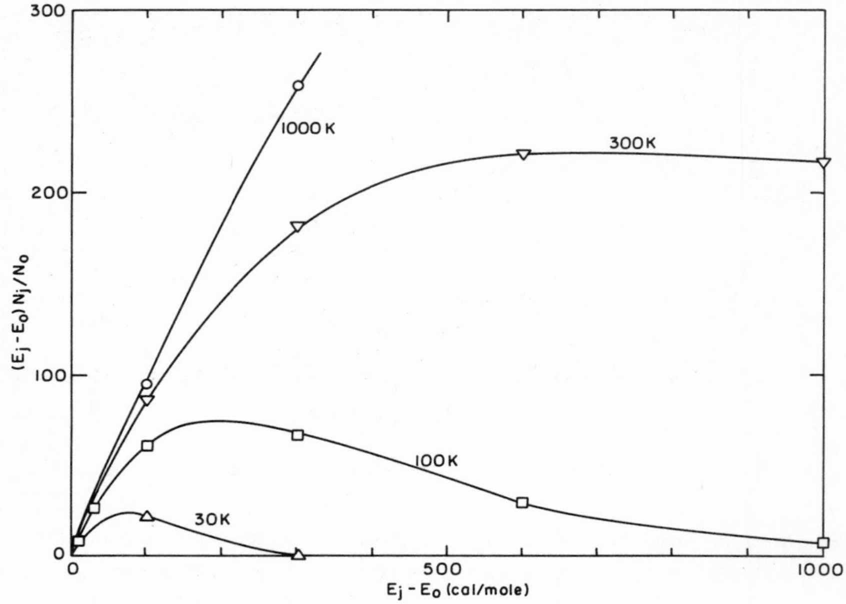

Fig. 4.6. Distribution of Energy

for instance, a molecule may have, in a higher vibrational state, a larger average moment of inertia than in a lower state. But such interactions need be taken into account only in a detailed examination. In general applications, assumption of independent energy sets

$$\varepsilon'_0,\ \varepsilon'_1\ \ldots\ \varepsilon'_j\ \ldots\ ; \qquad \varepsilon''_0,\ \varepsilon''_1\ \ldots\ \varepsilon''_k$$

is permissible. The energy in any class is then given by

$$\varepsilon_{jk} = \varepsilon'_j + \varepsilon''_k; \qquad u_{jk} = u'_j + u''_k\ . \tag{4.42}$$

The repetition of the preceding discussion results then in Boltzmann's theorem

$$N_{jk} = N_{oo}\, e^{-u_{jk}} \tag{4.43}$$

and in multiplicative combinations

$$g_{jk} = g'_j \cdot g''_k; \qquad Q = Q' \cdot Q''\ . \tag{4.44}$$

The thermodynamic functions, depending on $\ell n\ Q$, are additive as would be expected. The generalization to more than two contributions is obvious.

This brief comment justifies the separate discussion of various contributions to the thermodynamic functions.

4.4 MOLECULAR VIBRATIONS

The vibrations of atoms in a molecule are of interest because they make a significant and peculiar contribution to the heat capacity and other properties; moreover, they offer a rewarding example of the useful application of molecular theory.

The essential features are well described by a simple mechanical model: A molecule is an assembly of atoms held together by forces that are sufficiently strong to prevent the escape of an atom. But the forces need not be as strong as for instance in a rigid glass, in which the motion of an atom against the others is reduced to a minimum.

The motions of an n-atomic molecule can be represented by the 3n components of the motion of its atoms. The same motions can be described and classified in a more significant manner. For a molecule in a dilute gas we lay the origin of a Cartesian coordinate system into the center of gravity of the molecule, and its axes into the principal axes of inertia. In this coordinate system the molecule as a whole carries out neither translational nor rotational motions. Motions of a part of a molecule, short of dissociation, cannot lead to an infinite distance; therefore they must be oscillations.

In order to describe the motion of the new coordinate system we need three spatial coordinates for the translation of the center of gravity, and in general three for the rotations. This leaves in general 3n-6 degrees of freedom for vibrational motions. Obvious exceptions are monoatomic molecules and linear molecules. If all molecules are arranged in a straight line (carbon dioxide) only two principal moments of inertia are different from zero and only two rotational motions take place. This leaves 3n-5 vibrations instead of 3n-6. Vibration diagrams for a few simple molecules are shown in Appendix 11.5.

Vibrations are characterized by their frequencies, which can be found in various kinds of spectra. It can happen that the frequencies of two or more different vibrations are equal, either approximately by accident, or precisely for some vibrations of certain molecules (which have a threefold or higher symmetry axis). Such vibrations are called degenerate. The number of <u>different frequencies</u> is then smaller than that of vibrations (Table 4.2).

Table 4.2

Motions of Gaseous Molecules

number of atoms	examples	arrangement	mechanical degrees of freedom transl.	rot.	vibr.	diff.freq.
1	He, Hg		3	0	0	0
2	N_2, HCl	linear	3	2	1	1
3	H_2O, SO_2	planar	3	3	3	3
3	CO_2	linear	3	2	4	3
5	CH_4		3	3	9	4
7	SF_6		3	3	15	6

Since Bjerrum's fundamental work of 1914 molecular vibrations have been extensively and thoroughly investigated by numerous authors [42]. As long as the amplitude of a vibration is small the force pulling an atom back to its average position (restoring force) is proportional to its distance from the rest position. Such an oscillator is called harmonic. Amplitudes are in general of the order of a few 10^{-10} cm(0.01 Å); only hydrogen amplitudes are about one order of magnitude larger. Deviations from the harmonic approximation can be found by measurements of high accuracy. The following discussion is restricted to harmonic oscillators. With this restriction the various vibrations of a molecule can be examined independently and their energies are simply added (their partition functions are multiplied).

Planck suggested in 1899 that energy is radiated in quanta proportional to the vibrational frequency ν, the proportionality factor

$$h = \Delta E/\nu \tag{4.45}$$

being a universal constant. It is called Planck's constant. A few years later Einstein proposed that the energies of subsequent states of a harmonic oscillator differed by $h\nu$. It was later found by various investigations that the energies of the vibrational states are

$$\varepsilon_o = 0.5\ h\nu; \quad \varepsilon_1 = 1.5\ h\nu; \quad \ldots \quad \varepsilon_j = (j+0.5)h\nu. \tag{4.46}$$

For the description of molecular vibrations we divide in general the frequency (sec^{-1}) by the light velocity $c = 2.99793 \cdot 10^{10}$ cm/sec and use the wave number

$$\bar{\nu} = \nu/c \tag{4.47}$$

which is the reciprocal of the wave length (number of waves per cm).

Introducing

$$u = h\nu/kT = hc\bar{\nu}/kT \tag{4.48}$$

into the ratios u_j of (4.27) we obtain

$$u_1 = u; \quad u_2 = 2u; \quad \ldots \quad u_j = ju \tag{4.49}$$

for the harmonic oscillator. The statistical weights g_j are unity for all states of an individual vibration.

Einstein's assumption furnishes at once the thermodynamic properties of an oscillator. We deduce from (4.28)

$$\begin{aligned} Q &= 1 + e^{-u} + \ldots e^{-u_j} + \ldots \\ &= 1 + e^{-u} + e^{-2u} + \ldots e^{-ju} + \ldots \\ &= 1/(1 - e^{-u}). \end{aligned} \tag{4.50}$$

The molal energy follows from (4.34)

$$\bar{E} - \bar{E}_o = RTue^{-u}/(1 - e^{-u}). \tag{4.51}$$

In order to find the molal heat capacity $\bar{C}_v$, we first change the last equation to

$$(\bar{E} - \bar{E}_o)/R = (hc\bar{\nu}/k)/(e^u - 1) \tag{4.52}$$

and differentiate with respect to T so that

$$\bar{C}_v/R = u^2e^{-u}/(1 - e^{-u})^2 . \tag{4.53}$$

This is the contribution of a single vibration. The total contribution is obtained by addition of the terms for each vibration.

Einstein's function (4.53) is shown in Fig. 4.7 for four vibrational wave numbers. Some molecular vibrations, mainly of heavy atoms such as bromine, are below 100 cm^{-1}; the diagram indicates that at room temperature the molal heat capacity is close to its limiting R, which we also can find for $u = 0$ from (4.53) by developing the exponential function into a series.

Only vibrations of hydrogen have wave numbers as high as about 3600 cm^{-1}. They do not contribute to the heat capacity at room temperature. For vibrations between 100 and 3000 cm^{-1} the contributions increase rapidly with the temperature.

The limiting value of the vibrational heat capacity at $T = 0$ is zero, as can be seen from the diagram of from (4.53). Moreover, all derivatives of $\bar{C}_v$ with respect to T have also the limit zero.

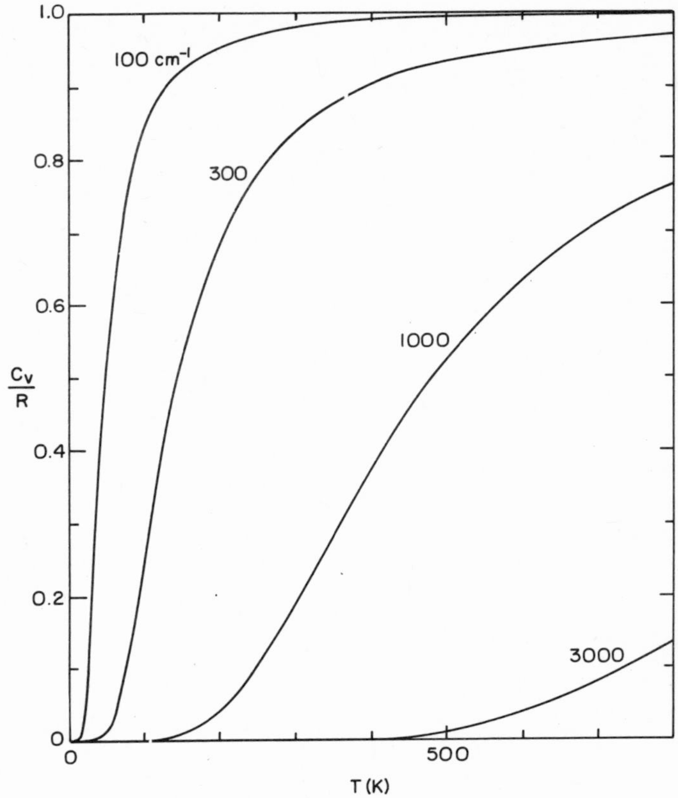

Fig. 4.7. Einstein Functions for the Heat Capacity (at various characteristic wave numbers)

The limiting value of $\bar{E}$ is of course $\bar{E}_o$ at T = 0 since all molecules are in the ground state. As the limit for high temperatures we find from (4.51) or (4.52) for u << 1 the result

$$\bar{E} - \bar{E}_o = RT . \qquad (4.54)$$

For a harmonic oscillator the average potential energy is equal to the average kinetic energy so that we have for high temperatures 0.5 RT for either.

The function (4.53) is in excellent agreement with measured heat capacities of gases. It does not very well fit the heat capacities of crystalline substances, for which it was originally derived. Debye took into account the interaction between molecules in a crystal and derived a similar but more complicated function. It is experimentally well confirmed. It differs from (4.53) in the approach to T = 0, giving

$$\bar{C}_V = DT^3 \tag{4.55}$$

as the limitng relation, with D being an individual parameter of the substance.

Tables of Einstein's and Debye's functions have been published in many places [43].

4.5 THE THIRD LAW

Early in this century the great importance of thermodynamic functions in the systematic description of chemical reactions was realized with increasing clarity. The practical advantage of reducing the need for experimental data by means of theoretical relations was well recognized. In this situation, a bold hypothesis was proposed by Nernst on a scanty experimental basis. He suggested that the entropy change ΔS approaches zero at T = 0 for any reaction between solids and liquids. The modification of Nernst's hypothesis by Planck, G. N. Lewis and others resulted in a principle of statistical thermodynamics, generally called the third law of thermodynamics.

The third law follows from the assumption that all molecules reach the same ground state (i.e., a state with the statistical weight $g_0 = 1$) when the temperature is sufficiently lowered. Under this assumption there exists only one microstate at zero temperature, and its thermodynamic probability is therefore W = 1. We conclude from Eq.(4.6) that the limit of the entropy equals the entropy constant $\bar{S}_0$,

$$\lim_{T=0} \bar{S} = \bar{S}_0 . \tag{4.56}$$

This is the lowest value of $\bar{S}$ since W = 1 is the lowest value of the thermodynamic probability. The third law says therefore that the constant $\bar{S}_0$ can be eliminated by extrapolation of entropy determinations extended to sufficiently low temperatures. Conventionally one sets $\bar{S}_0 = 0$ and calls "absolute" the entropy referred to the condition

$$\lim_{T=0} \bar{S} = 0 . \tag{4.57}$$

Equation (4.56) follows also from (4.38) since Q according to (4.29) is unity under the assumption of a single microstate.

The validity of this assumption has been thoroughly discussed by many authors. It is obvious from the temperature-independent equation (4.11) that it cannot hold for a perfect gas; but it has been pointed out that every gas condenses at a finite temperature so that this question is moot. Actually Bose as well as Einstein have shown that at extremely low temperatures even the kinetic energy is distributed in discrete quantum states and that for such a "degenerate gas state" the assumption holds.

The second case for questioning the assumption is a solution, frozen either as a solid solution or as a glass, i.e., as a liquid solution in which the molecules have fixed positions. In this case the entropy exceeds the entropy of the pure solids by the entropy of mixing. In first approximation it is given by the term (4.19) for the perfect solution. Whatever additional terms may be caused by molecular interactions, the entropy of mixing is certainly greater than zero. Even for pure substances in the state of a glass the entropy has positive values down to $T = 0$ because the molecules exist in states of different energies due to the influence of their neighbors. This has been shown experimentally by the measurement of the entropies from very low temperatures up to the temperature of liquefaction for glycerine [44] and iodic acid. Both substances can be obtained at low temperatures as glasses or (by proper seeding) as crystals. The determination of the heat capacities and the heat of liquefaction furnishes the entropy values (Fig.4.8).

It is true that a glass or a solid solution is not in a stable equilibrium with respect to the pure components. But equilibrium with respect to a small variation of some variable is all that is required in thermodynamic discussion.

A third case presents greater difficulty. If the energy difference $\bar{E} - \bar{E}_o$ of a state is only a few calories or a fraction or one calorie, the upper state will retain an appreciable population on cooling down to temperatures of the order of the characteristic temperature $(\bar{E} - \bar{E}_o)/R$. This may well be below the accessible temperature range (see Eq. 4.41). The correct application of the third law depends, however, on the extension of measurements below the temperature characteristic of the transition from the ground state to the lowest elevated state. Otherwise the zero level of entropy will be set at a temperature where two states are present with approximate populations of $N_1 = N_0 = 0.5$. The partition function then will be $Q = 2$, and the actual entropy $\bar{S} = R\ell n\ Q = 1.987 \cdot 2.303 \cdot 0.301 = 1.378$ cal/deg. mole.

A similar case is presented by carbon monoxide. The carbon and oxygen units of the molecule are so similar that a turn by 180° from CO to OC in

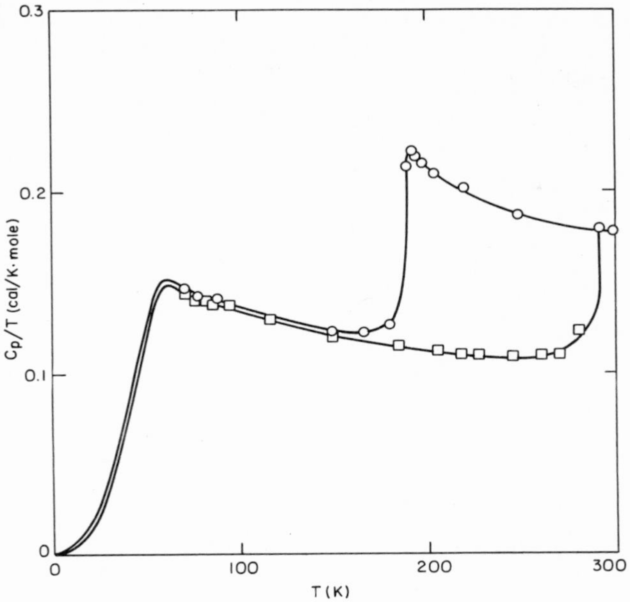

Fig. 4.8. Glycerin (the entropy of the glass is represented by the area under the upper curve, the entropy of the crystal by the area under the lower curve)

the crystal changes the energy of interaction with the neighbors only by an extremely small amount. A random distribution of the directions is therefore maintained on cooling to arbitrarily low temperatures. The entropy level of 1.0 cal/deg.mole, found [45] by extrapolation to T = 0, is a little lower than the value $R\ell n\ 2$ computed for complete randomness of the direction.

Simpler is the case of isotopes. For any reaction, in which the isotopic composition is not changed for any molecular species, terms of the kind $R\ell n\ Q$ appear equally on both sides of a reaction equation. They are therefore conventionally disregarded. But this cannot be done, of course, for reactions in which isotopic separations occur as for instance in various reactions of hydrogen and its compounds.

A survey of this discussion shows that the third law cannot be formulated in a manner similar to the first or second law. We may say that the entropy approaches zero at T = 0 if at the same time all molecules reach the same ground state. This would be a meager statement if it could not be amplified by the addition that the condition is satisfied as a rule by crystals of a pure substance. Exceptions are not numerous; sometimes they can be predicted, more often they can be explained on the basis of molecular theory. A solution (mixed crystal or glass) retains the entropy of mixing on cooling.

Thus the application of the third law depends on information in the field of molecular theory and on measurements of the heat capacity down to a temperature below $(\bar{E} - \bar{E}_0)/R$ for the lowest transition.

The practical importance of the third law nevertheless has been very great. It has been a keystone in building up systematic information of heat capacity, entropy, and free energy as functions of the temperature. This development is a good example for the mutual promotion of theoretical and experimental investigations.

4.6 HEAT CAPACITY AND ENTROPY

The information provided by molecular theory leads to a helpful classification of the various terms contributing to the heat capacity. They are surveyed for crystals and gases in Table 4.3.

Table 4.3
Contributions to the Molal Heat Capacity

	Molecules	Crystal	Gas
Translation	all	0	1.5 R
Rotation	monoatomic	0	0
	hydrogen	irregular	0 at low temperatures
	linear	0	R
	chain molecules	above transition	-
	spherical molecules	above transition	-
	polyatomic	-	1.5 R
Vibration		DT^3, Debye function	Einstein functions
Anharmonicity			bT
	free electrons	bT	
$\bar{C}_p - \bar{C}_v$		$T\frac{\partial P}{\partial T}\frac{\partial \bar{V}}{\partial T}$	(perfect: R)
Interaction		$\frac{\partial \bar{C}_p}{\partial P} = -T\frac{\partial^2 \bar{V}}{\partial T^2}$	$\frac{\partial \bar{C}_p}{\partial P} = -T\frac{\partial^2 \bar{V}}{\partial T^2}$

4.61 <u>Translation</u>. In a crystal the positions of the molecules are fixed so that there is no energy of translation.

In a gas the translational energy is 1.5 RT at all temperatures that have been experimentally realized. For still lower temperatures the theory of gas degeneracy predicts discrete quantum states and lower energies. The valence electrons in a metal are in a similar state.

The translational energy 1.5 RT leads to a heat capacity contribution

$$\bar{C}_V = (\partial\bar{E}/\partial T)_V = 1.5\ R \qquad (4.58)$$

and therefore according to (3.7) to an entropy contribution

$$\bar{S} = \int\bar{C}_V\ dT/T = 1.5\ R\ell n\ T + \text{const.} \qquad (4.59)$$

The constant cannot be determined by extending the integration to T = 0 and applying the third law, since (4.58) does not hold for the range of the degeneracy. But experimental data and theory, developed by Sackur, Tetrode and Stern, replaced this integration, demonstrated the dependence on the molal weight $\bar{M}$, and furnished the numerical value for the integration constant

$$\bar{S} = 1.5\ R\ell n\ T + R\ell n\ \bar{V} + 1.5\ R\ell n\ \bar{M} + 11.074\ \text{cal/deg.mole.} \qquad (4.60)$$

The dependence on the volume is prescribed by (4.10) for the perfect gas (volume in cc/mole).

This expression may be considered to be the "absolute" translational entropy of a perfect gas. In general, it is the total entropy of a monoatomic gas at low pressure. For other gases, other terms are to be added.

Replacing $\bar{V}$ in (4.60) by means of the equation for the perfect gas

$$\bar{V} = RT/P; \qquad R = 82.0562\ \text{cc.atm/deg.mole} \qquad (4.61)$$

we obtain for the translational entropy

$$\bar{S} = -R\ell n\ P + 2.5\ R\ell n\ T + 1.5\ R\ell n\ \bar{M} - 2.315\ \text{cal/deg.mole} \qquad (4.62)$$

with the pressure to be expressed in atmospheres (the gas constant in (4.60) and (4.62) is 1.987 cal/deg.mole).

4.62 <u>Rotation</u>. Rotational quanta are inversely proportional to the moment of inertia. For monoatomic molecules the moments of inertia are so low that the first energy state above the ground level cannot be reached at any realizable temperature. The same holds for the rotation of linear molecules around the molecular axis.

Other rotations are in crystals almost always inhibited by neighbor molecules. There are two exceptional classes.

Normal paraffins have enough available space to rotate around the molecular axis in a temperature range of a few degrees below their melting points. The onset of this rotation is observed by anomalous absorption of energy on heating. The observed heat capacity goes through a maximum and returns to a value slightly higher than that for the lower temperatures.

Similar transition points due to the start of rotations in crystals have also been observed far below the melting point in spherical molecules and ions such as methane and ammonium salts.

The moments of inertia of polyatomic molecules are so large and the rotational quanta therefore so low that in a gas always many rotational levels are occupied. The average kinetic energy is then 0.5 RT for every degree of freedom, i.e., 1.5 RT for a non-linear molecule. A linear molecule such as CO_2 rotates only around the two axes perpendicular to the molecule axis. The rotational contribution is therefore RT.

There is a single exception for a linear molecule, namely H_2. The combination of low mass and short distance results in a low moment of inertia. At room temperature rotations around the two axes perpendicular to the H-H axis are excited and the rotational contribution to $\bar{C}_V$ has the value R, normal for linear molecules. At lower temperatures the rotational contribution decreases. The quantitative description of the rotational heat capacity is complicated by the existence of two modifications of the hydrogen molecule, para-hydrogen and ortho-hydrogen.

Numerous moments of inertia have been deduced with high accuracy from spectra in which the lines are due to transitions from one rotational state to another one. From the moments of inertia the rotational contribution to energy, entropy and heat capacity can be computed.

4.63 Vibration. Energy steps in vibrations are in general considerably greater than those of rotations. It has been pointed out in Section 4.4 that only for low vibrations, say with wave numbers below 100 cm^{-1}, a close approach of the limiting value R is reached at room temperature (300 K). The highest frequencies (about 3600 cm^{-1}) contribute only at high temperatures.

The empirical rule of Dulong and Petit states that for a monoatomic crystal

$$\bar{C}_p = 6.4 \text{ cal/deg.mole.} \tag{4.63}$$

This is approximately equal to 3 R, the limiting value of $\bar{C}_v$ for vibrations of atoms in the three directions of space. In general, $\bar{C}_p$ is a little higher than $\bar{C}_v$. The rule holds well for low vibration frequencies, i.e., for soft and heavy elements such as lead (weak forces between atoms of high mass) but not at all for diamond.

The additivity of $\bar{C}_p$ (Kopp's rule) is a crude empirical relation for the solid state. Its validity range is about the same as that of the rule of Dulong and Petit. The heat capacity for an n-atomic substance can therefore be estimated as 6.4 n cal/deg.mole.

4.64 Other Contributions; Liquids. Small contributions can be caused by deviations of vibrations from the model of a harmonic oscillator. Similarly, the free electrons in a metal may gain translational energy. Both effects are linear in the temperature and usually negligible.

Interaction between molecules of a gas is expressed by the equation of state. Its influence on the heat capacity is given by Eq.(3.13).

The difference between $\bar{C}_p$ and $\bar{C}_v$ has been derived in (3.18).

We have little theoretical guidance for the heat capacity of liquids. It is in general higher than that of the same substance in the gaseous or crystalline state at the same temperature.

SUMMARY OF THE FOURTH CHAPTER

Statistical thermodynamics greatly contributes to our understanding by presenting plain and transparent models for some essential ideas of thermodynamics. Idealized substances, the perfect gas and the perfect solution, are models of outstanding usefulness. Molecular theory helps us also in describing moderate deviations from the behavior of these idealized substances. The classification and description of the thermal properties of gases and solids is largely based on molecular models.

The most important contribution of statistical theory to general thermodynamics is the interpretation of entropy by means of an enumeration of the states of the molecular model, appearing to our coarse senses as the same state. In statistical thermodynamics the zero point of temperature has a good meaning as a limit; under certain, well understood conditions, a pure crystal approaches a single molecular state at zero temperature. Thus a natural zero level of entropy is introduced.

References

(41) R. H. Fowler and E. A. Guggenheim, Statistical Thermodynamics, McMillan Co., New York, 1929. J. E. Mayer and M. G. Mayer, Statistical Mechanics, John Wiley & Sons, New York, 1940. M. Dole, Statistical Thermodynamics, Prentice-Hall, New York, 1954. F. Reif, Fundamentals of Statistical and Thermal Physics, McGraw-Hill, New York, 1965. The first two books are general standard works, the other two present clear discussions of the problems of special interest in chemical thermodynamics.

(42) G. Herzberg, Molecular Spectra and Molecular Structure, Vol. 1 (1944); Vol. 2 (1955), Van Nostrand Co., New York.

(43) Debye's functions are tabulated in Lewis-Randall-Pitzer-Brewer, Appendix 5.

(44) G. E. Gibson and W. F. Giauque, J. Am. Chem. Soc., 45 (1923) 93. F. Simon and E. Lange, Z. Physik, 38 (1926) 227.

(45) J. O. Clayton and W. F. Giauque, J. Am. Chem. Soc., 54 (1932) 2610.

CHAPTER 5: MATERIAL CHANGES

The main subject of chemical thermodynamics is the application of the general laws and experimental information discussed in the preceding chapters to material changes of any kind. These include chemical reactions in a single phase, equilibria between phases of the same substance, chemical reactions involving different phases, and also nuclear reactions. This chapter will develop the general technique and discuss pure phases; the subject of the following chapters will be the application to solutions.

5.1 COMPONENTS

The individual substances that together build up a composite object or "system" are called components. There is always a minimum number c of components, also called the number of independent components. Nitric oxide, oxygen and water, for instance, are sufficient to build up a system containing not only NO, O_2 and H_2O, but also NO_2, N_2O_3, N_2O_4, N_2O_5, HNO_2 and HNO_3. The same system can also be produced by NO, N_2O_5 and HNO_2 or by other combinations of components but not by fewer than three, the minimum number in this example.

The number c depends on the reactions admitted in the system. At high temperatures nitric oxide decomposes to some extent to nitrogen and oxygen; at room temperature, however, the reaction rate is so slow that nitrogen present must be considered to be a fourth independent component. The set of independent components is complete only if each phase individually can be built up. If calcium carbonate is partially decomposed to calcium oxide and carbon dioxide, the system has two independent components because the gas phase and the calcium oxide phase require two different components (which suffice to form the second solid phase, calcium carbonate). A single component (calcium carbonate) could build up the system only with the restriction that the two phases CaO and CO_2 are present in equal amounts.

As a rule, we characterize the amounts of the components present by their number of moles n_1, n_2, ... n_i, ... n_z. These quantities satisfy the conditions established for coordinates in Section 1.25.

In an equilibrium between two phases without a chemical reaction the condition of interaction, corresponding with Eq.(1.1), is given by the relation

$$dn_i' + dn_i'' = 0 \tag{5.1}$$

between the number of moles of component i in the two phases. For the equilibrium in a chemical reaction the interaction conditions are given by the stoichiometric relations. For the decomposition of calcium carbonate, for example, we set up the following scheme, distinguishing the three phases by primes:

$$CaCO_3(s) = CaO(s) + CO_2(g) \tag{5.2}$$

i	1	2	3
n_i' (first solid)	n_1'	0	0
n_i'' (second solid)	0	n_2''	0
n_i''' (gas)	0	0	n_3'''
ν_i	-1	1	1.

The reaction is subject to the interaction conditions

$$dn_1' + dn_2'' = 0 \tag{5.3}$$

$$dn_1' + dn_3''' = 0. \tag{5.4}$$

We introduce the reaction coefficients, small integers or fractional numbers proportional to the dn_i's, namely,

$$\nu_1 = -1; \qquad \nu_2 = 1; \qquad \nu_3 = 1. \tag{5.5}$$

For the dissociation of nitrogen tetroxide we write

$$N_2O_4 = 2NO_2 \tag{5.6}$$

i	1	2
n_i	n_1	n_2
ν_i	-1	2.

We make it a rule to use negative numbers for the reaction coefficients of the reactants, positive numbers for the products.

5.2 THE PHASE RULE

The distinction of extensive and intensive quantities leads to a very useful relation for heterogeneous equilibria, first noticed by Gibbs. At first we restrict the discussion to an equilibrium between c independent components and admit only chemical, thermal and pressure interaction. The equilibrium system then is characterized by c + 2 independent variables, e.g., the energy, the volume and the total numbers of moles of the independent components.

We call p the number of phases present in equilibrium. The amount of any phase certainly may be considered to be an independent variable since an increase or decrease of the amount of any phase can be carried out without disturbing the amounts of the other phases or the intensive properties of any phase. If we deduct these p <u>necessarily</u> extensive variables from the total number we obtain the "number of degrees of freedom"

$$f = c + 2 - p \qquad (5.7)$$

i.e., the number of independent variables which may be intensive <u>or</u> extensive. Any of these can be made intensive since we can change any extensive variable by dividing it through the amount of the corresponding phase. Therefore f variables are essentially intensive.

Whenever we are interested only in the nature of the phases considered but not in total amounts, the phase rule gives us the relevant number f of the independent intensive properties.

Any kind of interaction other than chemical, thermal or by pressure introduces another degree of freedom beyond the number f of Eq.(5.7).

As an example we consider a vessel containing water in the solid, liquid and gaseous state. For the equilibrium of one component in three phases the phase rule furnishes f = 0, i.e., no intensive quantity can change as long as all three phases are present (triple point in Fig. 5.1). On addition of heat, some ice melts but the temperature and the pressure, the specific gravity or the refractivity of any of the phases remains unchanged: The system is nonvariant (For this reason it can be fairly easily reproduced; it has been chosen for the definition of the thermodynamic temperature scale by the convention T = 273.1600 K at the triple point of water. The freezing point of water at atmospheric pressure is then 273.15 K. Read "K" as "degree Kelvin").

If enough heat is added to melt the ice, the phase rule gives f = 1, i.e., one intensive variable can be arbitrarily varied in the presence of

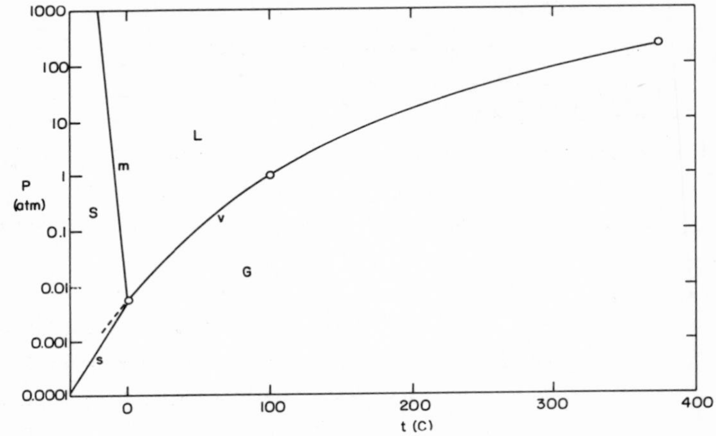

Fig. 5.1. Phase Diagram of Water (S ice, L water, G vapor; s sublimation curve, m melting curve, v vapor pressure curve. The broken lines indicates the vapor pressure of supercooled water)

two phases. The system is univariant and represented by a line v in Fig. 5.1. On addition of heat some liquid vaporizes, and the pressure and the temperature increase. But we can choose only <u>one</u> intensive property, say, the temperature; the pressure, called the vapor pressure or saturation pressure, is then fixed. It is a function of the temperature. Before we can decrease the pressure below the vapor pressure we have to vaporize the whole liquid so that we obtain a divariant system.

5.3 THE CHEMICAL POTENTIAL

Since the numbers of moles n_1, n_2 ... are generalized coordinates, we may write the equations (2.32), (2.33) and (2.38) for material changes as

$$dE = TdS - PdV + \Sigma\mu_i \, dn_i \tag{5.8}$$

$$dA = -SdT - PdV + \Sigma\mu_i \, dn_i \tag{5.9}$$

$$dG = -SdT + VdP + \Sigma\mu_i \, dn_i . \tag{5.10}$$

The quantity μ_i has been introduced in these equations as the generalized force conjugated to the mole number n_i. It has been called "chemical potential" by Gibbs. According to the discussion of Section 1.26 the chemical potential μ_i represents the force acting against an increase of the mole number n_i. According to the derivation of (2.32), (2.33) and (2.38), the same quantities μ_i appear in (5.8), (5.9) and (5.10). From these equations we conclude immediately that

$$\mu_i = \left(\frac{\partial E}{\partial n_i}\right)_{S,V} = \left(\frac{\partial A}{\partial n_i}\right)_{T,V} = \left(\frac{\partial G}{\partial n_i}\right)_{T,P} \qquad (5.11)$$

where all mole numbers except n_i are to be kept constant.

The significance of the chemical potential as a generalized force is illuminated by the isopiestic measurements of Robinson and Sinclair [51]. In order to determine the chemical potential of water μ_1'' in a solution of, say, sodium sulfate, they compare it with the chemical potential μ_1' in a solution of potassium chloride which had been established before by other methods. For the comparison one puts open weighing bottles with the various solutions in a vessel, evacuates it and submerges it in a thermostat. By weighing the bottles after appropriate periods of time and repeating the whole procedure with KCl-solutions of various concentrations one finds that KCl-solution which neither loses water to the unknown Na_2SO_4 solution nor gains any water from it. Then the chemical potentials μ_1' and μ_1'' are equal. The procedure is a precise chemical analog of the potentiometric measurement of a voltage or of the measurement of any other generalized force.

The equality of the chemical potentials in the isopiestic determination is equivalent with the equality of other quantities, e.g., the vapor pressures of water over the two solutions (the pressures at which pure gaseous water is in equilibrium with each of the solutions). G. N. Lewis coined the name "escaping tendency" for all functions which control a chemical equilibrium in the same way the chemical potential does. According to Section 1.13 this is the group of monotonically increasing functions of the chemical potential.

Since the most convenient potential function in problems of chemical thermodynamics is the thermodynamic potential or the free energy of Gibbs, we shall as a rule use the notation

$$\bar{G}_i = \left(\frac{\partial G}{\partial n_i}\right)_{T,P} \qquad (5.12)$$

for the chemical potential and also call it the partial molal free energy of Gibbs.

If non-chemical interactions are excluded, the condition of equilibrium (2.38) for a chemical reaction carried out at constant temperature and pressure becomes now

$$dG = \Sigma \bar{G}_i \, dn_i = 0. \qquad (5.13)$$

Introducing the reaction coefficients ν_i as in the scheme (5.2) and indicating by Δ a change specified in a reaction equation, we have now

$$\Delta G = \Sigma\nu_i\bar{G}_i = 0. \tag{5.14}$$

A reaction proceeds irreversibly if

$$\Delta G = \Sigma\nu_i\bar{G}_i < 0. \tag{5.15}$$

For the example introduced by (5.2) we obtain

$$\Delta G = \bar{G}_2'' + \bar{G}_3''' - \bar{G}_1' < 0. \tag{5.16}$$

The primes indicate the phases for which the partial molal quantities are to be derived. In this special example all three phases involved happen to be pure so that the partial molal quantities are equal to the molal quantities.

If any other kind of interaction takes place, as for instance in a galvanic element, the appropriate term has to be added on the way from (2.38) to (5.10). This case will be discussed in detail in Chapter 10.

The same terminology and notation is used for various quantities. The change in volume or heat content accompanying a specified chemical reaction can be represented by

$$\Delta V = \Sigma\nu_i\bar{V}_i \tag{5.17}$$

$$\Delta H = \Sigma\nu_i\bar{H}_i. \tag{5.18}$$

The measurement and computation of partial molal quantities will be discussed in detail in the following chapter.

Since the differentiation indicated in (5.12) is independent of temperature and pressure, we may derive from (5.12), (3.4), (3.8), (3.10)

$$\left(\frac{\partial\bar{G}_i}{\partial T}\right) = \frac{\partial}{\partial T}\left(\frac{\partial G}{\partial n_i}\right) = \frac{\partial}{\partial n_i}\left(\frac{\partial G}{\partial T}\right) = -\frac{\partial S}{\partial n_i} = -\bar{S}_i \tag{5.19}$$

$$\frac{\partial}{\partial T}\left(\frac{\bar{G}_i}{T}\right) = -\frac{\bar{H}_i}{T^2}; \qquad \frac{\partial\bar{G}_i}{\partial P} = \bar{V}_i. \tag{5.20}$$

In general, the symbol Δ will be used for a change proceeding at constant temperature and pressure. Any deviation from this rule should be clearly stated.

5.4 STANDARD STATES

Although the principles of chemical thermodynamics had been known much earlier, efficient calculation methods were introduced only by G. N. Lewis in this century. An important point in the success of the new methods has been the systematic introduction of standard states.

Since energy and entropy are defined as integral functions only differences of these functions for two states have a direct significance. The systematic recording of data and their application becomes clumsy if in addition to the state of interest another must always be dragged along. For this reason, Lewis and later authors introduced a set of rules defining standard or reference states to which energy, entropy and free energy data are referred. In final applications, which always involve differences between two states, the standard states cancel out provided that the same standard states are chosen.

The rules offer the great practical advantage that in general no attention need be payed to the standard states. Exceptional cases, in which special explanations are required or deviations from the rules are desirable, do not very often occur.

For the entropy the third law suggests a simple rule: The entropy of a pure crystal at the absolute zero temperature is zero. The apparent and real exceptions and supplements have been discussed in Sections 4.5 and 4.6. Starting from this basis, we can establish the entropy $\bar{S}^{\circ}$ of a pure substance at atmospheric pressure as a function of the temperature by means of calorimetric measurements, integrating $\bar{C}_p/T$ and adding this term and the entropy changes for phase transitions. This temperature function $\bar{S}^{\circ}$ is, at any given temperature, the standard value for the measurement or computation of entropies at any pressure and for their dependence on the composition of solutions.

In practical calculations the heat content is usually preferred to the energy. The basic convention adopts the value zero for the heat content of each element in its stable state at a fixed temperature. Most tables are based on the temperature 298.15 K for the simple reason that data for lower temperatures are usually less accurate or missing. But some tables are based on the fixed temperature 0 K; they are restricted to those substances for which sufficient information for a reasonable extrapolation to zero temperature is available.

The basic heat content of a pure substance is thus the increase of heat content on formation from the elements at 1 atm, either $\bar{H}^f_0$ at 0 K or $\bar{H}^f_{298}$ at 298.15 K. The standard heat content $\bar{H}^{\circ}$ at one atm for any temperature is obtained from either value by adding the integral of $\bar{C}_p$ and the enthalpy changes for phase transitions.

For water, for instance, we find in the tables of Pitzer and Brewer [52]

$$H_2(g) + 0.5\ O_2(g) = H_2O(g); \qquad \Delta H = \bar{H}^f_o = -57107 \text{ cal}$$

$$H_2(g) + 0.5\ O_2(g) = H_2O(g); \qquad \Delta H = \bar{H}^f_{298} = -57798 \text{ cal}$$

i: 1 2 3

The two enthalpy changes are represented by

$$\bar{H}^f_o = \bar{H}^o_o(3) - \bar{H}^o_o(1) - 0.5\bar{H}^o_o(2) \tag{5.21}$$

$$\bar{H}^f_{298} = \bar{H}^o_{298}(3) - \bar{H}^o_{298}(1) - 0.5\bar{H}^o_{298}(2). \tag{5.22}$$

The difference between the two standard enthalpies of formation is

$$\bar{H}^f_{298} - \bar{H}^f_o = [\bar{H}^o_{298}(3) - \bar{H}^o_o(3)] - [\bar{H}^o_{298}(1) - \bar{H}^o_o(1)] - 0.5[\bar{H}^o_{298}(2) - \bar{H}^o_o(2)]. \tag{5.23}$$

In the same tables we find the corresponding numerical values

$$\bar{H}^f_{298} - \bar{H}^f_o = 2368 - 2024 - 0.5 \cdot 2070 = -691. \tag{5.24}$$

Consequently the data based on 0 K and on 298 K are different; in the application to any reaction the difference, of course, cancels out <u>if the same basis is used</u>. The difference is <u>not</u> given by the first bracket term in Eq.(5.23), as one might assume on superficial inspection.

From $\bar{H}^o - \bar{H}^f_o$ or $\bar{H}^o - \bar{H}^f_{298}$ and $\bar{S}^o$ one derives the numerical values

$$\bar{G}^o - \bar{H}^f_o = \bar{H}^o - \bar{H}^f_o - T\bar{S}^o ;$$

$$\bar{G}^o - \bar{H}^f_{298} = \bar{H}^o - \bar{H}^f_{298} - T\bar{S}^o. \tag{5.25}$$

For any reaction one derives from appropriate tabulated data

$$\Delta G^o = \Delta(G^o - H^f_{298}) + \Delta H^f_{298} . \tag{5.26}$$

While $\bar{G}^o$, being an integral function, has no definite value, the differences $\bar{G}^o - \bar{H}^f_{298}$ and ΔG^o are well defined.

Some tables [53] contain the values of $\bar{G}^f = \Delta G^o$ for the formation of a compound from the elements. The change of the free energy for a reaction between any compounds can then be obtained simply as

$$\Delta G^o = \Delta G^f \tag{5.27}$$

since the terms referring to the elements cancel out.

While a discussion of standard states is necessary and their introduction is practically inevitable, they should be taken as a tool and not be cultivated as an end for themselves. For simplicity and clarity standard states should be introduced only sparingly and eliminated as soon as possible.

5.5 PHASE EQUILIBRIA

5.51 A Single Substance. The phase rule (5.7) tells us that a single component in a single phase has two degrees of freedom, i.e., that all intensive properties are functions of two of them. An equation of state, for instance, expresses the molal volume as a function of temperature and pressure.

In the presence of two phases the pressure can be represented as a function of the temperature. The discussion of the shift of this equilibrium with the temperature will be specialized for liquid water and steam. The equilibrium

$$H_2O(\ell) = H_2O(g); \qquad \Delta G = \bar{G}_g - \bar{G}_\ell = 0 \qquad (5.28)$$

exists along the saturation line v between the liquid and gaseous fields in Fig. 5.1. If we differentiate (5.28) along this line we obtain in view of (2.38)

$$\left(\frac{d\Delta G}{dT}\right)_{sat} = -\Delta S + \Delta V \cdot \left(\frac{dP}{dT}\right)_{sat}$$

$$= -\bar{S}_g + \bar{S}_\ell + (\bar{V}_g - \bar{V}_\ell) \cdot dp/dT = 0. \qquad (5.29)$$

We call the equilibrium pressure in this case vapor pressure or saturation pressure and denote it by p. Its temperature dependence is given by Clapeyron's equation

$$dp/dT = \Delta S/\Delta V = \Delta H/(T\Delta V) \qquad (5.30)$$

where we write

$$\Delta H = \bar{H}_g - \bar{H}_\ell; \qquad \Delta V = \bar{V}_g - \bar{V}_\ell. \qquad (5.31)$$

Since $\Delta G = 0$, the increase in heat content ΔH on vaporization equals $T\Delta S$.

Equation (5.30) can be derived in the same way for any univariant equilibrium of any substance, in particular, for any of the curves in Fig. 5.1. We see immediately that dp/dT is positive if ΔH and ΔV have the same sign. For the melting curve of water dp/dT is negative because water contracts on melting so that ΔV is negative for the melting process; ice melts on compression, as shown immediately by the diagram.

5.52 Vapor Pressure Equations. The relation between vapor pressure and temperature, as a practical problem, has several features in common with the equation of state: both are important for application, for both numerous relations have been suggested, and different relations are useful for different purposes.

Any equation of state that covers both the liquid and gaseous state implies a vapor pressure relation. But existing equations of state have not been good enough to encourage much development work in this direction. The usual starting point is Clapeyron's equation (5.30); various assumptions for ΔH and ΔV for the vaporization provide a series of relations simultaneously increasing in refinement, number of individual parameters, and adaptability to widely different substances.

The simplest assumptions are independence of temperature for ΔH, the perfect gas equation for V_g, and a negligibly small value for V_ℓ as compared with V_g. On this basis Eq.(5.30) leads to the Clausius-Clapeyron equation

$$\frac{dp}{dT} = \frac{\Delta H}{RV_g} = p\frac{\Delta H}{RT^2} \qquad (5.32)$$

$$\log p = A - B/T; \qquad B = \Delta H/(2.303\ R). \qquad (5.33)$$

At the atmospheric boiling temperature the calculated value of (5.32) is a few percent too low, but the equation is very useful for interpolation at low and moderate pressures with empirical coefficients.

In the next step a linear temperature function

$$\Delta H = \Delta H_o + T\Delta C_p \qquad (5.34)$$

is introduced for the heat of vaporization in (5.32). The resulting equation of Rankine

$$\log p = A - B/T - C \log T; \qquad B = \Delta H_o/(2.303\ R); \quad C = -\Delta C_p/T \qquad (5.35)$$

has been often used in earlier times. Another relation with three parameters, the empirical relation of Antoine,

$$\log p = A - B/(C + t); \qquad t = T - 273.15 \qquad (5.36)$$

is a little more convenient and just as adaptable to experimental data.

These equations have been extensively used up to pressures of 1 atm and slightly above. They fail halfway between the boiling point and the critical temperature. An inflection point, demonstrated by experimental data (Fig.5.2), cannot be represented by Antoine's equation, and not with acceptable parameters by Rankine's equation.

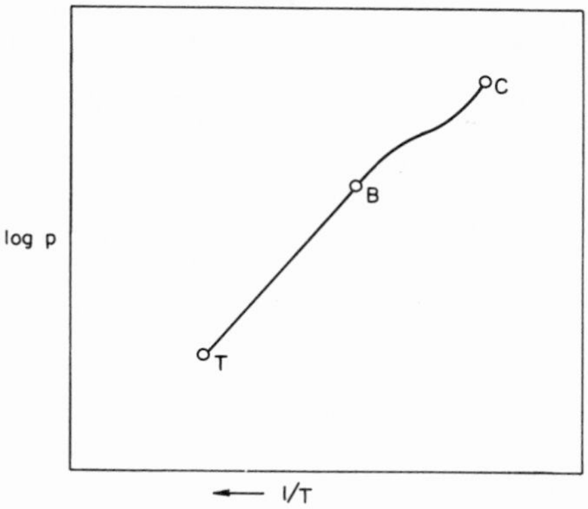

Fig. 5.2. Vapor Pressure p (t triple point, b boiling point, c critical point. The curvature is greatly exaggerated)

A thorough study of the experimental data by Thodos [54] led Frost and Kalkwarf [55] to the relation

$$\log p \;=\; A - B/T - C \log T + Dp/T^2. \tag{5.37}$$

The last term is the result of a very reasonable estimate of the influence of term $\overline{V}_g - \overline{V}_\ell$, which deviates from RT/p rapidly on approach on the critical point. The equation indeed furnishes a good representation of the vapor pressure in the whole range of the liquid, thus balancing the inconvenience of having p on the right hand side.

These equations appear to offer a sufficient choice for practical demands.

5.53 Reactions Between Pure Phases. Univariant equilibria between pure phases are similar to phase equilibria of a pure substance. If one (and only one) substance is present in the gaseous state, relations of the type of vapor pressure equations are useful.

A discussion of how to apply the principles developed so far to a concrete example will be preferable to an abstract derivation. We shall continue the examination of the thermal decomposition of calcium carbonate begun with Eqs.(5.2) and (5.16), using the notation introduced there. As before, we indicate the standard state for the Gibbs free energy, i.e., the value for 1 atm at any temperature T, by a superscript °. The primes are unnecessary here since the phase of the standard state is fixed by convention.

For the solid phases we derive G_j as a function of pressure from Eq.(3.10) so that

$$CaCO_3(s): \quad \bar{G}_1' = \bar{G}_1^o + \int_1^P \bar{V}_1 dP = \bar{G}_1^o + \bar{V}_1^o(P-1) \tag{5.38}$$

$$CaO(s): \quad \bar{G}_2'' = \bar{G}_2^o + \int_1^P \bar{V}_2 dP = \bar{G}_2^o + \bar{V}_2^o(P-1). \tag{5.39}$$

For moderate pressure we may approximate the molal volumes by their standard values; for low pressures (a few atmospheres) and moderate accuracy, we may even drop the integral terms (often called Poynting terms) entirely.

For the gaseous phase we have from Eq.(3.47)

$$CO_2(g): \quad \bar{G}_3''' = \bar{G}_3^o + RT\ln(P\phi_3). \tag{5.40}$$

The fugacity coefficient ϕ_3 approaches unity at low pressures.

The last three equations furnish for the change in Gibbs free energy the relation

$$\Delta G = \bar{G}_2'' + \bar{G}_3''' - \bar{G}_1' = \Delta G^o + RT\ln(P\phi_3) + (\bar{V}_2^o - \bar{V}_1^o)(P - 1) \tag{5.41}$$

for moderate pressures, or

$$\Delta G = \Delta G^o + RT\ln P \tag{5.42}$$

for low pressures. The standard ΔG^o of the Gibbs free energy can be computed from the data given for the three substances in available tables. It is shown as a function of 1/T in Fig. 5.3.

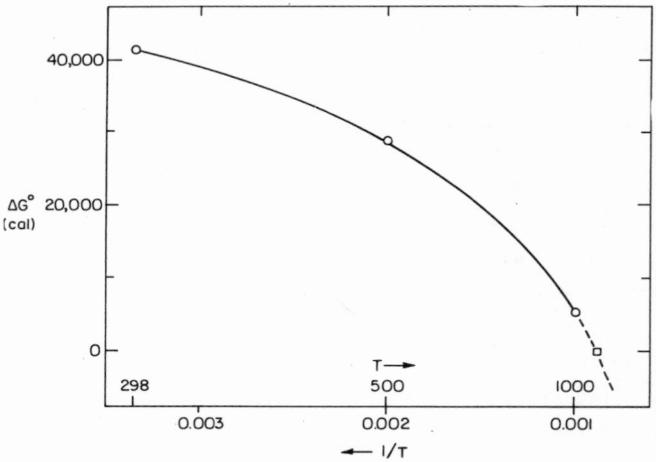

Fig. 5.3. Decomposition of Calcium Carbonate (o experimental data, □ extrapolated decomposition point at atmospheric pressure)

For equilibrium we write p for the pressure (decomposition pressure). Since ΔG is zero at equilibrium, we obtain from (5.42) for low pressures

$$\log p = -\Delta G^\circ/(2.303\ RT) = -\Delta G^\circ/(4.5758\ T). \qquad (5.43)$$

The results are shown in Fig. 5.4.

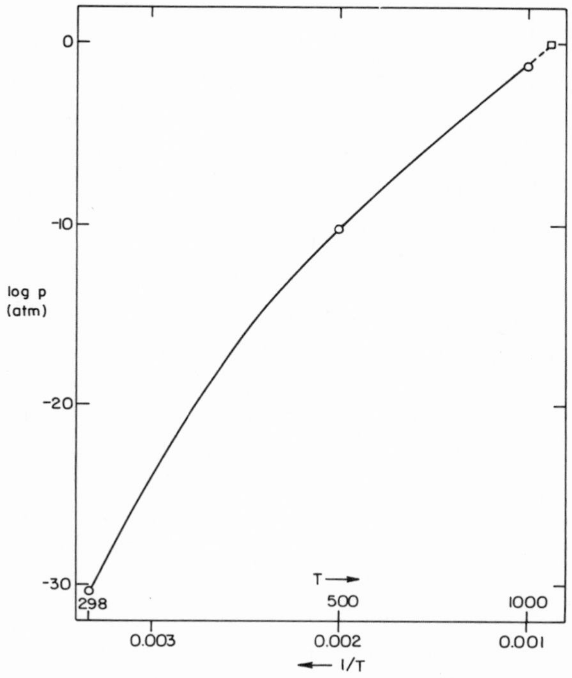

Fig. 5.4. Decomposition Pressure of Calcium Carbonate

Whenever the actual pressure P of the pure carbon dioxide phase is lower than the equilibrium pressure, the expression (5.42) for ΔG becomes negative and the decomposition of calcium carbonate proceeds spontaneously. If P is greater than p, the process runs spontaneously in the opposite direction.

A slight extrapolation in Fig. 5.3 or Fig. 5.4 shows that the decomposition temperature of calcium carbonate under atmospheric pressure [56] is 1150 K. The high stability at room temperature is indicated by the value $p = 4.05 \times 10^{-30}$ atm, which we compute from the thermodynamic data. This value is of course far below the level at which a pressure has an immediate significance, but it can perfectly well be used in the application to related problems.

The range covered by the decomposition pressure is much larger than that of any vapor pressure because the temperature coefficient is determined by the change in heat content, which is very often much greater for a chemical reaction than for any vaporization. In addition, a greater temperature range is represented in Figs. 5.3 and 5.4 than that accessible to a liquid.

SUMMARY OF THE FIFTH CHAPTER

Material changes are chemical reactions or the transfer of a substance from one phase to another.

The distinction of extensive and intensive quantities in heterogeneous equilibria leads to the practically important phase rule of Gibbs.

The number of moles of a substance in a phase is a generalized coordinate. The conjugated generalized force is the chemical potential, which is also the partial molal free energy of Gibbs.

The integration constants with which energy and entropy are affected are disposed of by rules for the choice of standard or reference states.

For the quantitative description of the vapor pressure as a function of the temperature one can choose from a set of efficient functions. All univariant equilibria between pure phases share some features.

References

(51) R. A. Robinson and D. A. Sinclair, J. Amer. Chem. Soc., 56 (1934) 1830. D. A. Sinclair, J. Phys. Chem., 37 (1933) 495.

(52) Lewis-Randall-Pitzer-Brewer, Thermodynamics, Appendix 7. McGraw-Hill, New York (1961).

(53) C. E. Wicks and F. E. Block, Thermodynamic Properties, Bulletin 605, Bureau of Mines, U. S. Government Printing Office, Washington (1963).

(54) G. Thodos, Ind. Eng. Chem., 42 (1950) 1514.

(55) A. A. Frost and D. R. Kalkward, J. Chem. Phys., 21 (1953) 264.

(56) Most of the numerous direct measurements are between 1160 and 1180 K (Gmelin's Handbook of Inorganic Chemistry, 8th Ed. Verlag Chemie, Weinheim, Germany, 1961. Vol. Calcium, 28B, p. 858). The agreement is satisfactory in view of the uncertainty of the extrapolation.

CHAPTER 6. SOLUTIONS

There is a special reason for the importance of solutions in the investigation as well as technical application of chemical thermodynamics. It is the wide range covered by the partial molal entropy $\bar{S}_j$ of a component j of a solution as a function of the composition. The range of $\bar{S}_j$ entails a similarly large range of the chemical potential $\bar{G}_j$ and therefore also of the change in Gibbs free energy ΔG of any reaction in which the substance j participates.

Conversely, a change in the composition of a solution is an efficient means of influencing the change of free energy. Such a change is a useful tool for measuring free energies and for guiding a chemical reaction in a desired direction.

Our main attention will be directed to the chemical potentials of gaseous mixtures and liquid solutions. Only the most general relations and results will be discussed here. A thorough discussion of the theory as well as applications can be found in a recent book by Prausnitz [61].

6.1 THE CHEMICAL POTENTIALS OF THE PERFECT SOLUTION

The relations for the perfect solution are so alluringly simple that they were often misused early in this century. It was G. N. Lewis who taught us to take full advantage of the simple rules and at the same time to describe imperfect solutions without any distortion. The device to be used for this purpose is the introduction of deviation terms.

In Eq.(4.19) we found that the entropy change in mixing x_1 moles of component A (molal entropy $\bar{S}_1$) and $x_2 = 1 - x_1$ moles of component B (entropy $\bar{S}_2$) to make a perfect solution at the same temperature and pressure is

$$\Delta S^* = \bar{S}^* - x_1\bar{S}_1 - x_2\bar{S}_2 = -R(x_1\ell n\, x_1 + x_2\ell n\, x_2) \tag{6.1}$$

so that the molal entropy of a perfect solution is

$$\bar{S}^* = x_1(\bar{S}_1 - R\ell n\, x_1) + x_2(\bar{S}_2 - R\ell n\, x_2). \tag{6.2}$$

For a perfect solution the change on mixing is zero for the heat content and the volume, as has been pointed out in Section 4.2. Heat content and volume are therefore linear functions of the values $\bar{H}_1^o$, $\bar{H}_2^o$, $\bar{V}_1^o$, $\bar{V}_2^o$ of the pure components:

$$\bar{H}^* = x_1\bar{H}_1^o + x_2\bar{H}_2^o; \qquad \bar{V}^* = x_1\bar{V}_1^o + x_2\bar{V}_2^o. \tag{6.3}$$

The partial molal entropy is derived from (6.2) by the prescription of Eq. (1.25). With

$$d\bar{S}^*/dx_1 = \bar{S}_1 - R\ell n\ x_1 - R - (\bar{S}_2 - R\ell n\ x_2) + R \tag{6.4}$$

we obtain

$$\bar{S}_1^* = \bar{S}^* + x_2 \cdot d\bar{S}^*/dx_1 = \bar{S}_1 - R\ell n\ x_1 \tag{6.5}$$

and, by exchanging subscripts,

$$\bar{S}_2^* = \bar{S}_2 - R\ell n\ x_2. \tag{6.6}$$

It is the logarithmic term in these relations that causes the large variation of $\bar{S}_1^*$ or $\bar{S}_2^*$ with the mole fraction.

From (6.3) we obtain for a perfect solution

$$\bar{H}_1^* = \bar{H}_1^o; \qquad \bar{H}_2^* = \bar{H}_2^o; \qquad \bar{V}_1^* = \bar{V}_1^o; \qquad \bar{V}_2^* = \bar{V}_2^o . \tag{6.7}$$

The partial molal free energy of Gibbs or the chemical potential follows from these relations as

$$\bar{G}_1^* = \bar{H}_1^* - T\bar{S}_1^* = \bar{H}_1^o - T\bar{S}_1 + RT\ell n\ x_1 = \bar{G}_1 + RT\ell n\ x_1 \tag{6.8}$$

$$\bar{G}_2^* = \bar{G}_2 + RT\ell n\ x_2. \tag{6.9}$$

These equations are valid for any pressure. We replace now the quantities $\bar{G}_1$ and $\bar{G}_2$ for the pure components by their standard values $\bar{G}_1^o$ and $\bar{G}_2^o$ using the relations (cf. Eq.(3.10))

$$\bar{G}_1 = \bar{G}_1^o + \int_1^P \bar{V}_1 dP; \qquad \bar{G}_2 = \bar{G}_2^o + \int_1^P \bar{V}_2 dP \tag{6.10}$$

so that we have instead of (6.8) and (6.9)

$$\bar{G}_1^* = \bar{G}_1^o + RT\ell n\ x_1 + \int_1^P \bar{V}_1 dP \tag{6.11}$$

$$\bar{G}_2^* = \bar{G}_2^o + RT\ell n\ x_2 + \int_1^P \bar{V}_2 dP. \tag{6.12}$$

One sees immediately that $\bar{G}_1^o$ is the value of $\bar{G}_1^*$ for $x_1 = 1$ and 1 atm. The partial molal free energy $\bar{G}_1^*$ (and likewise the partial molal entropy $\bar{S}_1^*$), however, is not defined for $x_1 = 0$ or $x_2 = 1$. This lack of definition expresses the fact that we cannot talk of a generalized force of component 1 to escape from a solution in component 2 if actually no component 1 is present. The situation is different for other quantities. The partial molal volume $\overline{V}_1$ has a limiting value for $x_1 = 0$; it is the change in the total volume on addition of 1 mole of component 1 to a large amount of component 2.

The results (6.11) and (6.12) can be extended to multicomponent solutions by the use of (1.37) and (4.22) instead of (1.25) and (4.19).

6.2 GASEOUS MIXTURES

6.21 Individual Fugacity Coefficients. In order to set up efficient calculation methods, we define the individual fugacity coefficient ϕ_j of component j by

$$\overline{G}_j = \overline{G}_j^o + RT\ell n(Py_j\phi_j). \tag{6.13}$$

For gases it is customary and convenient to write y for a mole fraction. Comparison of (6.13) with (6.11) furnishes

$$RT\ell n\ \phi_1 = \overline{G}_1 - \overline{G}_1^* + \int_1^P \overline{V}_1 dP - RT\ell n\ P. \tag{6.14}$$

For a perfect gas we have

$$\int_1^P \overline{V}_1 dP = RT\int_1^P \frac{dP}{P} = RT\ell n\ P. \tag{6.15}$$

The fugacity coefficient expresses therefore a combination of the deviation from the perfect solution $\overline{G}_1 - \overline{G}_1^*$ and of the deviation from the perfect gas.

We obtain the molal Gibbs free energy of a binary mixture according to Eq.(1.28) by multiplying (6.13) by y_1 and adding the corresponding term for component 2. Thus we find

$$\begin{aligned}\overline{G} = y_1\overline{G}_1 + y_2\overline{G}_2 &= y_1\overline{G}_1^o + y_2\overline{G}_2^o + RT(y_1\ell n\ y_1 + y_2\ell n\ y_2)\\ &+ RT\ell n\ P + RT(y_1\ell n\ \phi_1 + y_2\ell n\ \phi_2).\end{aligned} \tag{6.16}$$

We may identify in this relation the perfect-solution term

$$\overline{G}^* = y_1\overline{G}_1^o + y_2\overline{G}_2^o + RT(y_1\ell n\ y_1 + y_2\ell n\ y_2), \tag{6.17}$$

which is independent of the pressure, and the deviation from the perfect gas-solution, which is expressed by means of the geometric mean fugacity coefficient

$$\ell n\phi = y_1\ell n\ \phi_1 + y_2\ell n\ \phi_2. \tag{6.18}$$

Using a term introduced by Scatchard, we call "excess molal free energy" and "excess partial molal free energies" the quantities

$$\overline{G}^E = RT\ell n\ \phi; \qquad \overline{G}_1^E = RT\ell n\ \phi_1; \qquad \overline{G}_2^E = RT\ell n\ \phi_2. \tag{6.19}$$

All relations derived in Section 1.34 apply then to these excess quantities; in particular Eqs.(1.25), (1.27) and (1.29) lead to

$$\ln \phi_1 = \ln \phi + y_2 d\ln \phi/dy_1; \qquad \ln \phi_2 = \ln \phi - y_1 d\ln \phi/dy_1 \tag{6.20}$$

$$\ln(\phi_1/\phi_2) = d\ln \phi/dy_1. \tag{6.21}$$

We may express (6.16), (6.18) and (6.19) by

$$\bar{G} = \bar{G}^* + \bar{G}^E + RT\ln P = \bar{G}^* + RT\ln(P\phi) \tag{6.22}$$

and notice the identity of this relation with Eq.(3.47), except that the perfect solution term $\bar{G}^*$ assumes the role of the standard value $\bar{G}^\circ$. Both represent the molal free energy at the given temperature and at 1 atm. As long as the composition of the mixture is not changed, $\bar{G}^*$ remains constant just as $\bar{G}^\circ$ is for a pure substance.

We may conclude that the determination of the mean fugacity coefficient of a mixture proceeds essentially in the same manner as the determination of the fugacity coefficient of a pure substance. No other method is in practical use. Individual fugacity coefficients are obtained from ϕ according to Eqs. (1.25), (1.27) and (6.19).

6.22 <u>The Equation of State for Mixtures</u>. The various equations of state, discussed in Section 3.3, apply to a mixture just as well as to a pure substance except for one important point: The critical point of a mixture has a different significance because at subcritical temperatures the mixture splits into two phases of <u>different</u> compositions. The critical temperature of a mixture is always higher than the temperature obtained by linear interpolation between the pure components. In view of the different meaning of the critical point it is not surprising that reduced equations of state cannot be directly used for mixtures.

According to a suggestion of W. B. Kay one uses the "pseudocritical" values of T_{pc}, P_{pc}, $\bar{V}_{pc}$, linearly interpolated between the critical values T_{cj}, P_{cj}, V_{cj} of the components,

$$T_{pc} = \sum_j y_j T_{cj}; \qquad P_{pc} = \sum_j y_j P_{cj}; \qquad \bar{V}_{pc} = \sum_j y_j \bar{V}_{cj} \tag{6.23}$$

for the reduction of the variables in an equation of state. The results are often satisfactory, particularly if the mixture consists of closely similar substances (e.g., paraffins).

The general problem is the calculation of the parameters, for instance a and b in Eq.(3.25) and (3.26), from the parameters a_j and b_j of the components. The best answer to this often discussed problem is the combination

$$a = (\Sigma y_j a_j^{0.5})^2; \qquad b = y_j b_j . \qquad (6.24)$$

Since a_j and b_j can be expressed by the critical constants (see Eqs.(3.20) and (3.30)), any combination rule implies a pseudoreduction rule to replace Kay's rule (6.23). For instance, rule (6.24) applied to Eq.(3.26) implies

$$T_{pc} = (\Sigma y_j T_{cj}^{1.25}/P_{cj}^{0.5})^{1.333}/(\Sigma y_j T_{cj}/P_{cj})^{0.667}$$

$$P_{pc} = (\Sigma y_j T_{cj}^{1.25}/P_{cj}^{0.5})^{1.333}/(\Sigma y_j T_{cj}/P_{cj})^{1.667}. \qquad (6.25)$$

As an example of the degree of approximation obtained by Eq.(3.26) and the simple combination rule, Fig. 6.1 shows the compressibility factors of n-butane-carbon dioxide mixtures at 311.0 K. The calculated values are based only on the critical temperatures and pressures of the two components.

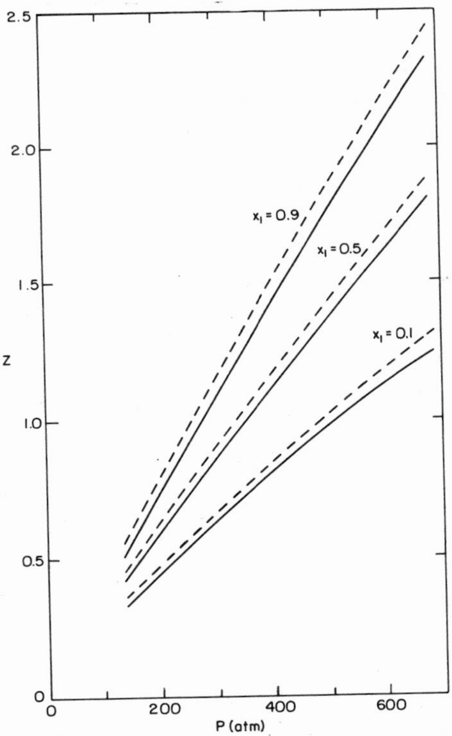

Fig. 6.1. Compressibility Factor Z of n-Butane-Carbon Dioxide (311.0 K. The full lines represent data of G. H. Sage and W. N. Lacey, "Some Properties of the Lighter Hydrocarbons", American Petroleum Institute, New York, 1955. The broken lines are calculated according to Amagat's rule)

Considerable improvement can be obtained by the introduction of specific interaction coefficients in a more general combination rule [62].

The essential advantage of an algebraic equation of state rests on the better calculation of fugacity coefficients. The performance of an integration (3.50)

$$\ln \phi = \int_{P^o}^{P} (Z-1)dP/P \tag{6.26}$$

and still more of a differentiation, as required in (6.20), inevitably entails a serious loss of accuracy if based on tabulated data or on diagrams.

For multicomponent mixtures we derive from (1.37) the individual fugacity coefficients

$$\ln \phi_i = \ln \phi + \frac{\partial \ln \phi}{\partial y_i} - \sum_a y_a \frac{\partial \ln \phi}{\partial y_a} \quad . \tag{6.27}$$

Introducing (6.24) into Eqs.(3.26) and (3.53) of Redlich and Kwong we obtain

$$\ln \phi_i = (Z-1)\frac{b_i}{b} - \ln\left(Z - \frac{bP}{RT}\right)$$

$$- \frac{a}{bRT^{1.5}}\left[2\left(\frac{a_i}{a}\right)^{0.5} - \frac{b_i}{b}\right]\ln\left(1 + \frac{bP}{RTZ}\right) \quad . \tag{6.28}$$

This relation turns into Eq.(3.53) for $y_i = 1$; $a_i = a$; $b_i = b$.

The simplest equation of state for gaseous mixtures is expressed in the rule of Lewis and Randall. It says that the fugacity coefficients are independent of the composition of the mixture and therefore equal to the fugacity coefficients of the pure components at the same temperature and pressure. Differentiation of Eq.(6.19) with respect to the pressure at constant composition shows immediately that the rule implies that the partial molal volumes $\bar{V}_1$ and $\bar{V}_2$ are also independent of the composition and equal to $\bar{V}_1^o$ and $\bar{V}_2^o$, respectively, of the pure components. At the same time the molal volume of the mixture must be a linear function of the mole fraction. This rule for the volume had been proposed long ago by Amagat.

The significant content of the rule of Lewis and Randall can be expressed as the statement: There is a pressure range in which a gaseous mixture is a perfect solution though not a perfect gas. This statement is a good approximation for some mixtures, especially of chemically similar substances, but it is not generally true. The rule, however, is a convenient tool for a crude estimation of fugacity coefficients and therefore frequently applied.

An example is shown in Fig. 6.1.

6.3 LIQUID SOLUTIONS

The procedure for liquid solutions starts out similarly to that for gaseous mixtures. Actually it is simpler because the pressure dependence of the free energy is small. When Lewis introduced the concept of the activity coefficient he left its application to gases open. It is advisable to restrict it to liquids and solids. Conversely the introduction of fugacity as a term for liquids is less advisable.

We discuss in this section non-dissociating substances; electrolytes will be the subject of Chapter 8.

6.31 The Activity Coefficient. The activity coefficient γ_j of component j in a liquid (or solid) solution is defined by

$$\bar{G}_j = \bar{G}_j^o + RT\ell n(x_j\gamma_j) + \int_1^P \bar{V}_j dP. \qquad (6.29)$$

In this definition the entire pressure variation of $\bar{G}_j$ is taken up by the integral. As a rule, we retain the convention that $\bar{G}_j^o$ is the molal Gibbs free energy of the pure substance j at the given temperature and at 1 atm. This convention implies that $\gamma_j = 1$ for the same standard state. Both $\bar{G}_j^o$ and γ_j are thus defined as independent of the pressure (this choice is not adopted by some authors).

The quantity

$$x_j\gamma_j = a_j \qquad (6.30)$$

is called the activity. The excess chemical potential of component j is

$$\bar{G}_j^E = RT\ell n\, \gamma_j. \qquad (6.31)$$

We introduce the excess molal free energy of the solution by means of Eq. (1.28) as

$$\bar{G}^E = RT\sum_j x_j \ell n\, \gamma_j \qquad (6.32)$$

so that the total molal free energy is

$$\bar{G} = \sum x_j\bar{G}_j^o + RT\sum x_j\ell n\, x_j + \bar{G}^E + \int_1^P \bar{V}dP \qquad (6.33)$$

with the molal volume according to (1.28) being $\bar{V} = \Sigma x_j\bar{V}_j$.

The reader will realize that these relations essentially are restatements of Sections 1.34 or 1.35. All relations derived there for partial molal quantities apply therefore also to $RT\ell n\, \gamma_j$. The relation of Gibbs, Duhem and Margules (Eq.(1.42)) can therefore be written as

$$\sum_j x_j \partial \ell n\, \gamma_j / \partial x_j = 0. \tag{6.34}$$

Since the value of the integral, called Poynting's term, is small (except in the vicinity of the critical point), the replacement of $\bar{V}_j$ by $\bar{V}_j^o$, the molal volume of the pure component j at 1 atm, is usually permissible. Practically it rarely matters whether we extend the integral to 1 atm or zero pressure. If we are not aiming at highest accuracy we may therefore write for (6.29)

$$\bar{G}_j = \bar{G}_j^o + RT\ell n(x_j\gamma_j) + P\bar{V}_j^o \tag{6.35}$$

and even omit the last term entirely for pressures of a few atmospheres.

For some calculations it is convenient to introduce the quantity

$$Q = \bar{G}^E/(2.303\ RT) = \sum_j x_j \log \gamma_j. \tag{6.36}$$

For binary solutions we write (see Eqs.(1.28),(1.25),(1.27) and (1.29))

$$Q = x_1 \log \gamma_1 + x_2 \log \gamma_2 \tag{6.37}$$

$$\log \gamma_1 = Q + x_2 \cdot dQ/dx_1; \qquad \log \gamma_2 = Q - x_1 \cdot dQ/dx_1 \tag{6.38}$$

$$\log(\gamma_1/\gamma_2) = dQ/dx_1. \tag{6.39}$$

These functions are useful in the examination and representation of experimental data, in the discussion of theorectical models and in applications. They are shown for ethanol-methylcyclohexane in Figs. 6.2, 6.3 and 6.4.

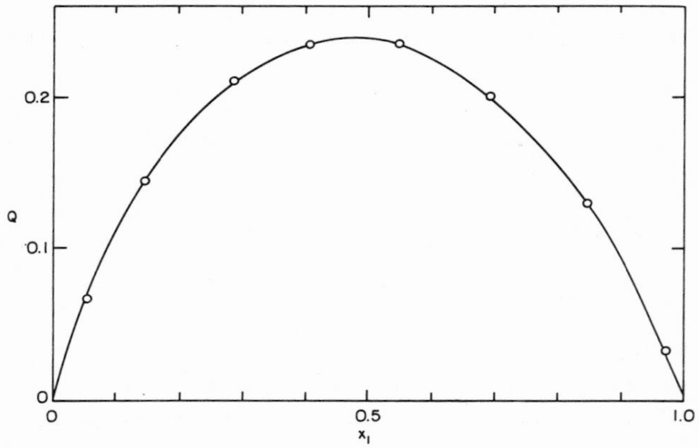

Fig. 6.2. Ethanol-Methylcyclohexane (35°C; data of C.B. Kretschmer and R. Wiebe, J. Am. Chem. Soc. 71 (1949) 3176. N. Isii, J. Soc. Chem. Ind. Japan 38, Suppl. (1925) 659

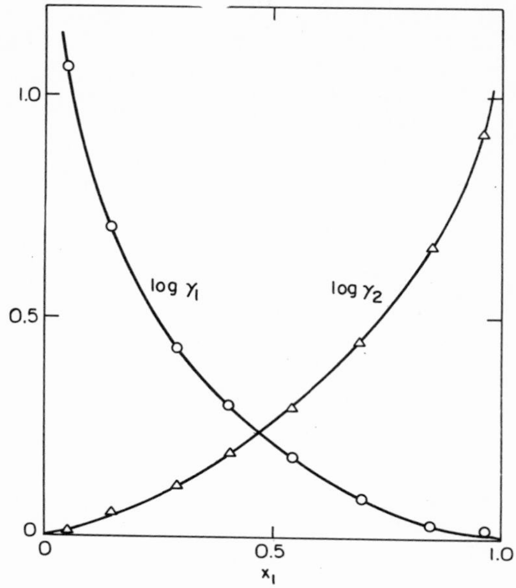

Fig. 6.3. Ethanol-Methylcyclohexane (activity coefficients; o $\log \gamma_1$, Δ $\log \gamma_2$)

The activity coefficients, representing deviations from the perfect solution, are often close to unity, sometimes large but always finite. The function Q (Fig. 6.2) therefore is zero for $x_1 = 0$ and $x_1 = 1$. Indeed, for $x_1 = 0$ or the pure second component, the first term in (6.37) is zero because $\log \gamma_1$ is finite, and the second term because $\gamma_2 = 1$ for the pure substance. The same holds for $x_1 = 1$ with changed subscripts. The limiting slopes of Q are according to (6.39)

$$\lim_{x_1=0} dQ/dx_1 = \lim_{x_1=0} \log \gamma_1; \qquad \lim_{x_1=1} dQ/dx_1 = -\lim_{x_1=1} \log \gamma_2 \qquad (6.40)$$

because $\gamma_2 = 1$ in the first case and $\gamma_1 = 1$ in the second case.

Integration of (6.39) furnishes [63.1,63.2] the <u>area condition</u>

$$\int_0^1 \log(\gamma_1/\gamma_2)dx_1 = 0. \qquad (6.41)$$

The integral is zero because Q has the value zero at the endpoints. This simple relation was useful for checking the reliability of experimental results at a time when accurate methods were being developed and the accumulation of extensive data had started. An example is shown in Fig. 6.5. The results of three observers agreed within rather large limits of scattering

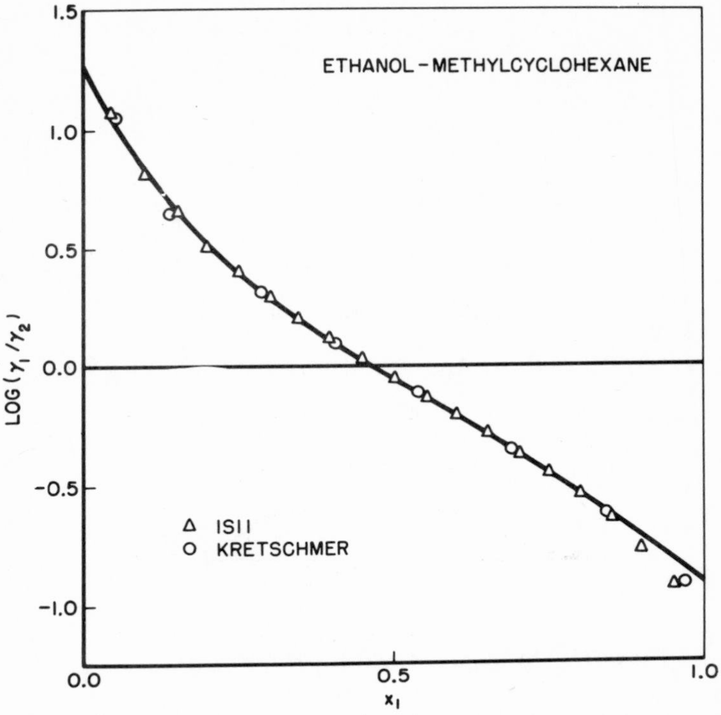

Fig. 6.4. Ethanol-Methylcyclohexane [$\log(\gamma_1/\gamma_2)$]. This figure and Fig. 6.6 have been taken from an earlier paper (Chem. Eng. Progr. Symp. Ser. 48, No. 2 (1952) 49)

but were incompatible with the area condition. The fourth set appears to be correct since the area under the straight line is zero as prescribed by (6.41).

From Eqs.(3.8) and (6.31) we conclude that

$$\left(\frac{\partial \ln \gamma_j}{\partial T}\right)_{P,x} = \frac{1}{R}\frac{\partial(\bar{G}_j^E/T)}{T} = -\frac{\bar{H}_j^E}{RT^2} \tag{6.42}$$

$$\bar{H}^E = -T^2\frac{\partial(\bar{G}^E/T)}{\partial T} \tag{6.43}$$

where $\bar{G}^E$ is given by (6.32). The change in heat content on mixing determines the temperature coefficient of the activity coefficient.

6.32 Algebraic Representation. Even before automatic computation became common, the representation of activity coefficients by suitable equations was highly desired for the purpose of checking and smoothing experi-

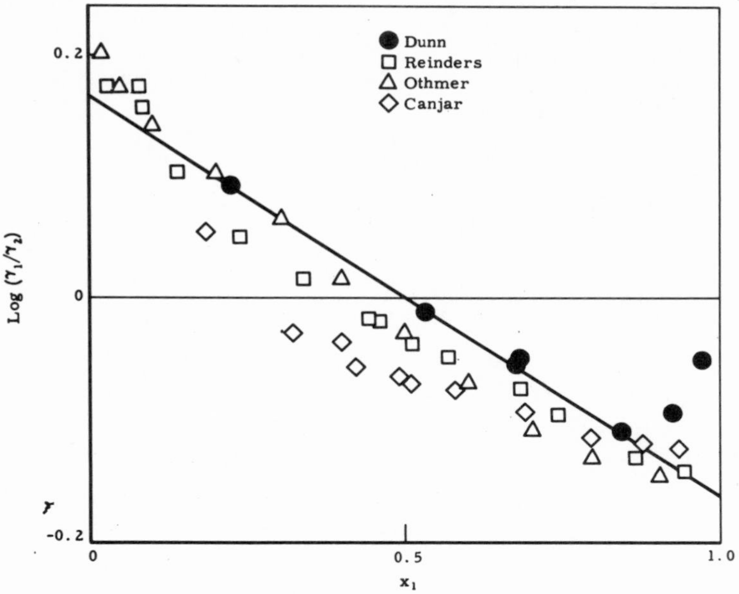

Fig. 6.5. Acetone-Benzene (• C. L. Dunn and G. J. Pierotti, Shell Development Company. □ W. Reinders and C. H. de Minjer, Rec. Trav. Chim. Pays-Bas 59 (1940) 369; △ D. F. Othmer, Ind. Eng. Chem. 35 (1943) 619; ◇ J. A. Tallmadge and L. N. Canjar, ibid. 46 (1954) 1279)

mental data, for converting them to different temperatures and pressures, and for extending them to other substances by reasonable theories or systematic guesses.

Margules [63.3] suggested the first and most general algebraic representation, namely, by a power series in the mole fraction. The most useful form for thermodynamic discussion and practical application [63.1] starts from Q. It must satisfy the zero values of Q for the pure components. For a binary solution one introduces

$$Q = x_1x_2[B + C(x_1-x_2) + D(x_1-x_2)^2 + \ldots]. \tag{6.44}$$

This representation offers some technical advantages. The coefficients B, D, F ... are unchanged if the subscripts are exchanged; the coefficients C, E ... only change signs. The number of terms used in each special case offers a reasonable distinction of various classes of solutions. A disadvantage of the power series is the lack of a theoretical foundation.

If all coefficients are zero we have a perfect solution, namely,

$$Q = 0; \quad \gamma_1 = 1; \quad \gamma_2 = 1 \tag{6.45}$$

in the whole concentration range.

The deviation of second degree from the perfect solution

$$Q = Bx_1x_2; \quad dQ/dx_1 = \log(\gamma_1/\gamma_2) = B(x_2-x_1) \tag{6.46}$$

$$\log \gamma_1 = Bx_2^2; \quad \log \gamma_2 = Bx_1^2 \tag{6.47}$$

is common to all representations as an approximation for small deviations from the perfect solution. This approximation represents Q by a parabola, symmetric in x_1 and x_2, and $\log(\gamma_1/\gamma_2)$ as a straight line. The example shown in Fig. 6.2 and 6.3 is far from satisfying these requirements. But frequently these relations are satisfactory, particularly if the components are chemically similar and not much different in molal volume. Sometimes the relations hold also for mixtures of chemically different components (Fig. 6.5). They are convenient for the precise representation of nearly perfect solutions [63.4]. Table 6.1 shows results for binary solutions of some C_8-aromatics at the atmospheric boiling temperatures. Eq.(6.47) shows that B is the highest value attained by $\log \gamma_1$ or $\log \gamma_2$; the value of γ_1 or γ_2 for ethylbenzene - m-xylene therefore does not exceed 1.0083. In a difficult technical distillation problem even such a small deviation from the perfect solution may be of interest.

Table 6.1

Solutions of C_8-Aromatics

			B
Ethylbenzene	-	o-Xylene	0.0035
Ethylbenzene	-	m-Xylene	0.0036
Ethylbenzene	-	p-Xylene	0.0031
m-Xylene	-	o-Xylene	0.0021
p-Xylene	-	o-Xylene	0.0019
p-Xylene	-	m-Xylene	0.0003

In general, two coefficients, B and C, are required and sufficient for a good representation of activity coefficients. But if there is hydrogen bonding between the molecules of a component, such as dimerization in acids or association in alcohols, three terms are usually necessary; even they are not always sufficient. For three terms we have

$$Q = x_1x_2[B + C(x_1-x_2) + D(x_1-x_2)^2] \quad (6.48)$$

$$dQ/dx_1 = \log(\gamma_1/\gamma_2) = B(x_2-x_1) + C(-1+6x_1x_2) + D(x_2-x_1)(1-8x_1x_2) \quad (6.49)$$

$$\log \gamma_1 = x_2^2[B + C(3x_1-x_2) + D(x_2-x_1)(x_2-5x_1) \quad (6.50)$$

$$\log \gamma_2 = x_1^2[B + C(x_1-3x_2) + D(x_2-x_1)(5x_2-x_1)]. \quad (6.51)$$

It follows from (6.39) that $\log(\gamma_1/\gamma_2)$ is one degree lower in x_1 than Q is. Therefore if the terms with B and C are sufficient, $\log(\gamma_1/\gamma_2)$ is well represented by a parabola (Fig. 6.6). The two-term equation, often called Margules equation, is satisfactory for mixtures of hydrocarbons and numerous other solutions. But Fig. 6.4 shows immediately that a parabola cannot represent $\log(\gamma_1/\gamma_2)$ of ethanol-methylcyclohexane. This is the rule for mixtures of a highly associated component (alcohol) and a non-associating component.

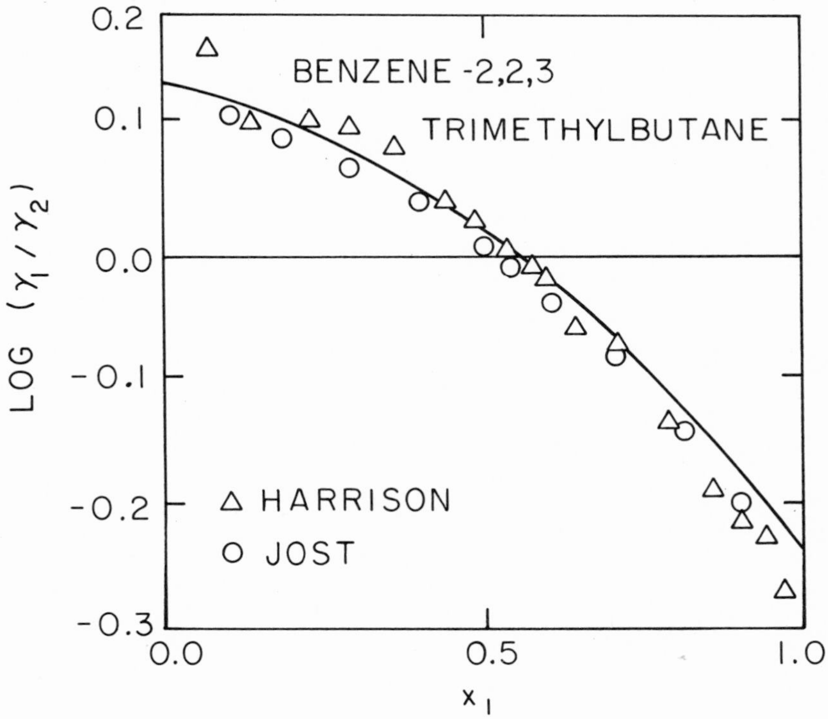

Fig. 6.6. Benzene-2,2,3 Trimethylpentane (J. M. Harrison and L. Berg, Ind. Eng. Chem. 38 (1946) 117; L. Sieg, Chem.-Ing.-Technik 1950, 322)

A power series such as (6.44) offers the advantage of a transparent, comprehensive survey of the behavior of all mixtures. Moreover, algebraic and numerical calculations are simple.

6.33 Van Laar's Equation. The problem of applying molecular theory, crude or sophisticated, to thermodynamic properties has always attracted considerable interest. The reasons are of course the better understanding attained and the high cost of experimental research. A good theory enables us to interpolate the available experimental data more efficiently and, moreover, to predict things unknown, thus saving time and money in experimental work.

Often used is the relation which has been called van Laar's equation

$$Q = G^E/(2.303\ RT) = A'B'x_1x_2/(A'x_1 + B'x_2) \tag{6.52}$$

(Actually van Laar's not very transparent derivations implied that the ratio A'/B' was the ratio of the critical volumes of the components). Eqs.(6.38) and (6.39) furnish

$$\log \gamma_1 = A'(B'x_2)^2/(A'x_1 + B'x_2)^2;$$
$$\log \gamma_2 = (A'x_1)^2B'/(A'x_1 + B'x_2)^2. \tag{6.53}$$

The values for $x_1 = 0;\ 0.5;\ 1$ are compared in Table 6.2. In the special "symmetric" case, namely, $A' = B'$ van Laar's equation is identical with the first term of the power series (6.44).

Table 6.2

Two-Parameter Equations
(Power series with B, C; van Laar's relation with A', B')

x_1		0	0.5	1
Q	Eq.(6.48)	0	0.25B	0
	Eq.(6.52)	0	$0.5A'B'/(A' + B')$	0
$\log \gamma_1$	Eq.(6.50)	B–C	$0.25(B + C)$	0
	Eq.(6.53)	A'	$A'(B')^2/(A' + B')^2$	0
$\log \gamma_2$	Eq.(6.51)	0	$0.25(B - C)$	B+C
	Eq.(6.53)	0	$(A')^2B'/(A' + B')^2$	B'
$\log(\gamma_1/\gamma_2)$	Eq.(6.49)	B–C	0.5C	–B–C
		A'	B'/A'	–B'

The equation of van Laar is a little less flexible than a power series with two coefficients. It does not allow a maximum for log γ_1. But it is as useful as the other relation and has the advantage of being related with crude theoretical interpretations [63.5].

6.34 The Relation of Scatchard and Hildebrand. The real and definitive answer to any questions regarding the thermodynamics of solutions would be offered by a function such as van Laar's function (6.52) that does not contain any empirical parameters but only quantities given by properties of the pure components; then the behavior of any solution can be really predicted.

In a way, this problem has been solved [63.6,63.7] independently by Scatchard and by Hildebrand. Such a result can be obtained, of course, only on the basis of several far reaching assumptions. The entropy of mixing is presupposed to be perfect (Eq.(4.19)) and deviations are considered only in the energy or heat content; such a solution has been called "regular" by Hildebrand.

The estimate of the energy per cm^3 required to remove a molecule from a solution is assumed to be the sum of terms proportional to the volume fractions of the components. Moreover, the energy required to remove a molecule X from a molecule Y is assumed to be the mean of the energies required for removing X from X and Y from Y. Both assumptions cannot be more than crude approximation since they disregard the influence of any third molecule on the interaction between two, and also disregard any influence of the structure or shape of either molecule. But they lead to a reasonable estimate of the change in heat content on mixing and thus of the function Q.

The partial molal volume $\overline{V}_j$ of the component j of a liquid mixture varies very little with the concentration (except for electrolyte mixtures) and differs not much from the volume $\overline{V}_j^\circ$ of the pure component. In the following we omit therefore the superscript and consider $\overline{V}_j = \overline{V}_j^\circ$ as independent of the composition of the solution. The total volume of one mole of solution is then approximately

$$\overline{V} = x_1\overline{V}_1 + x_2\overline{V}_2 \tag{6.54}$$

and the volume fractions are

$$v_1 = x_1\overline{V}_1/\overline{V}; \qquad v_2 = 1 - v_1 = x_2\overline{V}_2/\overline{V}. \tag{6.55}$$

The energy of vaporization of one cm^3 of the pure components (Scatchard's "cohesive energy density" or the square of Hildebrand's "solubility parameter") is represented by δ_1^2 and δ_2^2, respectively. We abbreviate the interaction term as

$$A = (\delta_1 - \delta_2)^2/(2.303\ RT). \tag{6.56}$$

The result of Scatchard and Hildebrand can now be derived [cf. 61] and expressed as

$$Q = A\bar{V}_1\bar{V}_2x_1x_2/\bar{V} = A\bar{V}v_1v_2 \tag{6.57}$$

$$\log \gamma_1 = A\bar{V}_1v_2^2; \qquad \log \gamma_2 = A\bar{V}_2v_1^2 . \tag{6.58}$$

These relations represent a special case of (6.52) and (6.53); the van Laar coefficients are now predicted as

$$A' = \bar{V}_1(\delta_1 - \delta_2)^2; \qquad B' = \bar{V}_2(\delta_1 - \delta_2)^2. \tag{6.59}$$

Thus, in principle, a complete answer to the prediction problem has been obtained.

Actually these relations often furnish approximate results, useful when experimental data are lacking. The approximation is poor for polar and for associated substances. A higher accuracy is desired in many applications such as for instance distillation problems.

From (6.59) one obtains a partial result

$$A'/B' = \bar{V}_1/\bar{V}_2 \tag{6.60}$$

for the ratio of the van Laar coefficients. This result is well confirmed by data for hydrocarbon mixtures but not for other solutions.

The relation of Scatchard and Hildebrand can be expanded for multicomponent solutions to

$$Q = 0.5\sum_{i,j}{}' A_{ij}\bar{V}_i\bar{V}_jx_ix_j/\bar{V} \tag{6.61}$$

with

$$\bar{V} = \sum_k x_k\bar{V}_k . \tag{6.62}$$

The accent at the summation symbol means that the summation is carried out over pairs of different subscripts, $i \neq j$.

One obtains van Laar's relation for multicomponent mixtures by divesting the quantities $\bar{V}_i$ of their meaning and considering them to be empirical parameters.

6.35 Entropy Deviations. The rules of the perfect solution were derived in Section 4.2 from two assumptions: equality of interaction between like and unlike molecules and equality of molecule volumes. The relaxation of the first assumption led to the interaction term of Scatchard and Hildebrand;

the elimination of the second assumption, which for a long time had not been given any attention, led to an important entropy term. It has been derived from various assumptions by Huggins [63.8] and by Flory [63.9] in 1940. Their remarkable result was the replacement of the perfect entropy of mixing (Eq.(4.19)) by

$$\Delta S_{FH} \;=\; -R(x_1 \ln v_1 + x_2 \ln v_2) \qquad (6.63)$$

i.e., the mole fractions in the logarithms are replaced by the volume fractions.

The excess molal entropy of mixing results then as

$$\begin{aligned}\bar{S}^E \;&=\; -R(x_1 \ln v_1 + x_2 \ln v_2) + R(x_1 \ln x_1 + x_2 \ln x_2) \qquad (6.64)\\ &=\; -R(\ln \bar{V} - x_1 \ln \bar{V}_1 - x_2 \ln \bar{V}_2)\end{aligned}$$

if we make use of the definitions of the preceding section and Eqs.(6.54) and (6.55). In view of the definitions (6.31) and (2.35) the contribution made to the activity coefficient γ_j by $\bar{S}^E$ is $-\bar{S}^E_j/R$. Together with an interaction term assumed to be given by (6.58) one obtains

$$\ln \gamma_1 \;=\; \ln(\bar{V}_1/\bar{V}) + (1 - \bar{V}_1/\bar{V}_2)v_2 + \mu v_2^2 . \qquad (6.65)$$

The notation

$$\mu \;=\; 2.303\, A\bar{V}_1 \qquad (6.66)$$

is conventional. For the activity $x_1\gamma_1$ one has

$$\ln(x_1\gamma_1) \;=\; \ln v_1 + (1 - \bar{V}_1/\bar{V}_2)v_2 + \mu v_2^2 . \qquad (6.67)$$

Disregarding for the moment the interaction term μv_2^2, we appreciate the magnitude of the Flory-Huggins effect by inspecting the limiting value γ_1^o for $v_2 = 1$; here the deviation from unity is largest. From Fig. 6.7 we see that the effect is small as long as the volumes are of the same order of magnitude (for $\bar{V}_2/\bar{V}_1 = 2$ we find $\log \gamma_1^o = -0.084$; $\gamma_1^o = 0.824$). But for large volume ratios the effect is considerable; it is therefore important for the solutions of a polymer in an ordinary solvent. Conventionally the subscript 1 is assigned to the solvent so that $\bar{V}_2/\bar{V}_1 > 1$.

Interaction effects usually (in the Scatchard-Hildebrand approximation always) contribute a positive term to $\log \gamma_1$; the Flory-Huggins term is always negative, as can be seen from Fig. 6.7 or derived from Eq.(6.63). For mixtures of small molecules the interaction effects prevail and a distinction of the two effects rarely can be ascertained. This is different for polymer solutions. The example of polyethylene solutions in n-heptane (Fig. 6.8) shows that here the Flory-Huggins effect prevails. The experimental curve goes sharply down to negative values and closer inspection shows that it cannot be represented by a parabola as would be expected for an interaction effect. However, log γ_1 can be very well represented by a Flory-Huggins term with the observed volume ratio $\bar{V}_2/\bar{V}_1 = 196$ and a moderate interaction term, namely, $0.150\ v_2^2$.

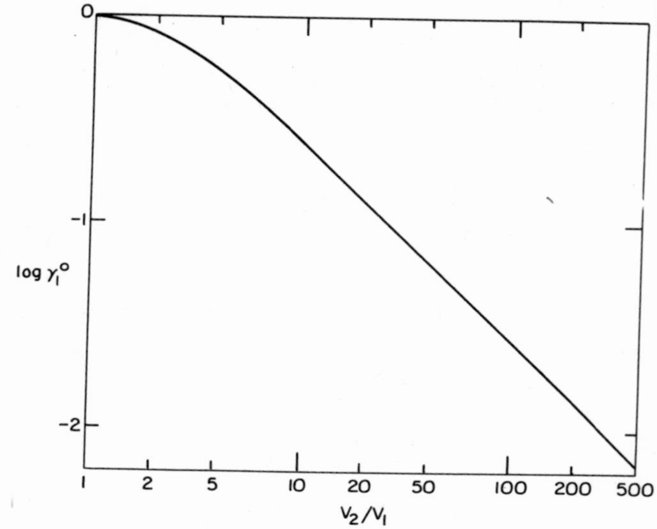

Fig. 6.7. Limiting value γ_1^o of the Activity Coefficient (for $v_2=1$, as obtained by the relation of Flory and Huggins as a function of the volume ratio V_2/V_1)

The diagram also shows that log γ_1 varies very little while v_2 goes up to 0.2.

The application of the relation of Flory and Huggins has been elaborately developed by Tompa, Koningsveld and other authors for applications in phase equilibria (solubility and precipitations) of polymer solutions, molal weight determinations, purification, and separation into fractions of uniform molal weight.

6.36 Association; Wilson's Function. The idea of chemical bonds between the molecules in a solution has been proposed surprisingly often and with considerable emphasis. It has always been an alluring prospect to retain nice and simple laws such as the relations of the perfect solution and to explain all deviations by the formation of compounds, solvation, association, ion pairs and so on. Numerous attempts of this kind, stubbornly maintained, failed completely.

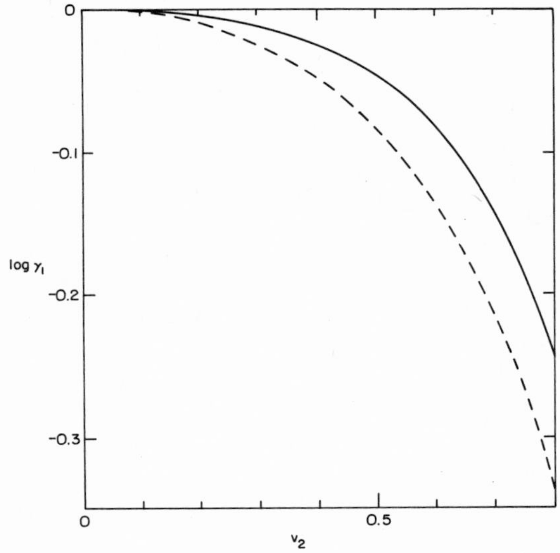

Fig. 6.8. n-Heptane-Polyethylene
(——— log γ_1 observed by J. H. van der Waals and J. J. Hermans, Rec. Trav. Chim. Pays-Bas 69 (1950) 971;
- - - Flory - Huggins term for V_2/V_1 = 196)

Sometimes such models are substantiated. Deviations from the perfect gas, found for acetic and a few other acids, have been quantitatively explained by Gibbs a century ago as due to an equilibrium of monomers and dimers. The anomalous maximum found for the activity coefficient of chloroform in ethanol solutions (Fig. 6.9) by Scatchard and Raymond indicates bonds between the components. In both cases the effect can be safely explained by hydrogen bonds.

More frequently the effect of hydrogen bonding is found in the self-association of alcohols and of a few other substances such as amines. A crude model [63.10] showed that for mixtures of associating and non-associating substances the curve for $\log(\gamma_1/\gamma_2)$ is S-shaped (see Fig. 6.4), clearly different from non-associating mixtures (Figs. 6.5, 6.6). The association effect can be crudely approximated by a term of fourth degree in (6.48), better by special functions [63.10,63.11] based on suitable assumptions. Association has been safely established by the measurement of properties of various kinds, such as infrared absorption, dielectric constants, nuclear magnetic resonance.

The characteristic effect of association on activity coefficients is reduced if both components are associating. Interassociation between

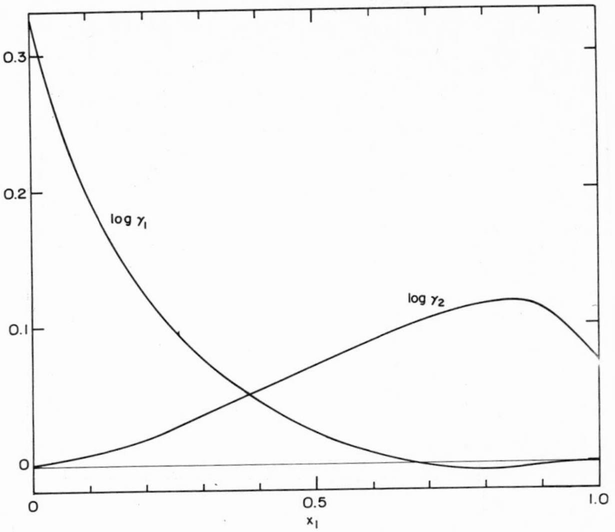

Fig. 6.9. Ethanol-Chloroform (G. Scatchard and C. L. Raymond, J. Am. Chem. Soc. 60 (1938) 1278.

components such as alcohols and water removes the S-shape of log (γ_1/γ_2).

An interaction function suggested by Wilson [63.12] successfully represents not only non-associating but also associating mixtures by means of only two parameters A" and B". For a binary solution Wilson suggests for the functions introduced in Eqs.(6.37) to (6.39)

$$2.303\ Q = -x_1 \ln(1-B''x_2) - x_2 \ln(1-A''x_1) \qquad (6.68)$$

$$\ln \gamma_1 = -\ln(1-B''x_2) + A''x_2^2/(1-A''x_1) - B''x_1x_2/(1-B''x_2) \qquad (6.69)$$

$$\ln \gamma_2 = -\ln(1-A''x_1) - A''x_1x_2/(1-A''x_1) + B''x_1^2/(1-B''x_2). \qquad (6.70)$$

These relations represent the activity coefficients of non-associating mixtures as well as the other two-parameter functions. Their great advantage is the excellent representation of associating solutions such as ethanol-methylcyclohexane which otherwise require a special association function containing a third parameter.

Wilson's function, however, requires a third parameter for the representation of incompletely miscible mixtures (Section 7.21). For this case Wilson multiplies the right-hand side of (6.68), and consequently of (6.69) and (6.70), by a constant C".

Wilson's relation can be extended to multicomponent mixtures without bias, it has the advantage that the binary parameters are sufficient to represent multicomponent systems.

The function has been examined by several investigators and parameters have been determined by now for a large number of mixtures [63.13].

The complexity of Wilson's function will be no real disadvantage in a routine procedure with a computer. For an occasional problem one will still take simpler functions into consideration.

SUMMARY OF THE SIXTH CHAPTER

The special importance of solutions in chemical thermodynamics results from the wide variation of the partial molal entropy and free energy with the concentration. Often we can make a reaction proceed in either direction or find a state of equilibrium simply by changing a concentration.

The free energy of a gaseous mixture is found in the same way as that of a pure gas, namely, by measuring the relation between pressure, volume and temperature. The concept of the fugacity coefficient is extended to describe the deviations from both the perfect gas and the perfect mixture.

It is convenient (though not generally accepted) to introduce the activity coefficient for the description of deviations from the perfect solution, and to restrict it to liquid and solid solutions, and to atmospheric pressure. The influence of pressure is separately taken into account.

A variety of functional relations between activity coefficients and concentrations is useful for the collection of information as well as for practical applications. Some of them are purely empirical, some constitute an improved knowledge of interaction between molecules, some describe the influence of the relative size of the molecules on their arrangement. For each specific purpose we have to select the most suitable one, taking into account performance and efficiency, the treasure of existing data, and convenience.

References

(61) J. M. Prausnitz, Molecular Thermodynamics of Fluid-Phase Equilibria, Englewood Cliffs, N. J.: Prentice-Hall, Inc., 1969.

(62) P. L. Chueh and J. M. Prausnitz, AIChE Journ. 13 (1967) 1099 and "Computer Calculations for High-Pressure Vapor-Liquid Equilibria"

Englewood Cliffs, N. J.: Prentice-Hall, Inc, 1968.

(63.1) O. Redlich and A. T. Kister, Ind. Eng. Chem. 40 (1948) 341, 345.

(63.2) E. F. G. Herington, Nature 160 (1947) 610.

(63.3) M. Margules, Sitzber. Akad. Wiss. Wien, Kl.II, 104 (1895) 1243.

(63.4) O. Redlich and A. T. Kister, J. Am. Chem. Soc. 71 (1949) 505.

(63.5) O. Redlich, E. L. Derr and G. J. Pierotti, J. Am. Chem. Soc. 81 (1959) 2283.

(63.6) G. Scatchard, Chem. Revs. 8 (1931) 321. J. Am. Chem. Soc. 56 (1934) 995. Trans. Faraday Soc. 33 (1937) 160.

(63.7) J. H. Hildebrand and S. E. Wood, J. Chem. Phys. 1 (1933) 817.

(63.8) M. L. Huggins, J. Chem. Phys. 9 (1941) 440. J. Phys. Chem. 52 (1948) 248.

(63.9) P. J. Flory, J. Chem. Phys. 9 (1941) 660. 10 (1942) 51.

(63.10) O. Redlich and A. T. Kister, J. Chem. Phys. 15 (1947) 849.

(63.11) G. Scatchard, Chem. Revs. 44 (1949) 7. I. Prigogine, V. Mathot and A. Desmyter, Bull. Soc. Chim. Belges 58 (1949) 547. P. J. Flory, J. Chem. Phys. 12 (1944) 425.

(63.12) G. M. Wilson, J. Am. Chem. Soc. 86 (1964) 127.

(63.13) J. M. Prausnitz, C. A. Eckert, R. V. Orye and J. P. O'Connell, Computer Calculations for Multicomponent Vapor-Liquid Equilibria, Prentice-Hall, Inc., Englewood Cliffs, N. J., 1967.

CHAPTER 7: PHASE EQUILIBRIA OF SOLUTIONS

For a gaseous mixture the determination of the free energy is fairly direct and simple: At sufficiently low pressures the law of the perfect gas mixture can be applied, and the results can be expanded to any pressure on the basis of the relation $\partial G/\partial P = V$ by means of P-V-T measurements.

No similar procedure can be used for liquid or solid solutions because only few mixtures are in sufficiently close agreement with the rules of the perfect solution. Several methods are used for determining the partial molal free energies or the activity coefficients; most are based on an equilibrium between two phases. If such an equilibrium exists with respect to the component j we have

$$\Delta G \;=\; \bar{G}_j - \bar{G}_j' \;=\; 0. \tag{7.1}$$

We denote, as a rule, the chemical potential of j in the liquid solution by $\bar{G}_j$, that in the other phase by $\bar{G}_j'$.

Such an equilibrium condition is applicable only if the component j is present in both phases; if a substance is not a component of a phase, its partial molal quantities cannot be defined (Section 1.34). For instance, in a vapor-liquid equilibrium of aqueous ethanol the condition applies to either component. But for a sugar solution and its vapor it applies only to water since sugar is not volatile. It also applies only to water in the equilibrium between a dilute sugar solution and ice. Similarly, Eq.(7.1) applies to water in an equilibrium established between a sugar solution and pure water through a semipermeable membrane, impermeable to sugar (in this case the two phases are under different pressures, the difference being called osmotic pressure).

The thermodynamics of liquid solutions rests on the investigation of phase equilibria. In addition, such equilibria are practically important because they control most technical methods of separating various substances. The efficiency of a certain phase equilibrium in the separation of component j (mole fractions x_j and y_j in the two phases) from component k is conventionally measured by the separation factor

$$\alpha_k^j \;=\; \frac{y_j}{y_k} \cdot \frac{x_k}{x_j} \,. \tag{7.2}$$

Its practical significance becomes obvious if we notice that in a flow process we may put in components j and k at the ratio x_j/x_k, and, at the same time take them out in the other phase at the ratio y_j/y_k. If $\alpha_k^j = 1$, the ratios of the mole fractions are equal in both phases so that the operation does not produce any separation. If α_k^j is large (or if it is much smaller than unity) the separation is efficient.

The search for the optimum conditions of separating by distillation, extraction, precipitation and so on, requires the knowledge of the separation factor as a function of the temperature, pressure and composition of the two phases.

In this chapter we shall concentrate our attention on non-dissociating substances.

7.1 VAPOR-LIQUID EQUILIBRIUM

7.11 Various Relations. In actual thermodynamic calculations some simplifying assumptions are often of great use. Sometimes the error introduced by these assumptions is below the influence of experimental errors, sometimes we must accept errors because the data are insufficient for a complete and exact computation. In every case we wish to find at least a rough estimate of the errors caused by calculation shortcuts.

In the following we shall first derive the exact relations and then discuss the various simplifying abbreviations. This procedure enables us to judge the advisability of abbreviations in each individual problem and protects us from the feeling of uncertainty that often results if one starts from the simplest formula and patches on correction terms. The introduction of fugacity and activity coefficients contributes a good deal to clarify the significance of simplifying assumptions.

We convert the equilibrium condition (7.1) for use in practical computation by introducing (6.29) and (6.13), putting a prime on properties of the gaseous phase. Thus we have

$$\bar{G}_j = \bar{G}_j^\circ + RT\ell n(x_j\gamma_j) + \int_1^P \bar{V}_j dP = \bar{G}_j' = \bar{G}_j^{\circ\prime} + RT\ell n(Py_j\phi_j). \qquad (7.3)$$

The standard values $\bar{G}_j^\circ$ and $\bar{G}_j^{\circ\prime}$ of component j in the two phases can be eliminated by means of the vapor pressure p_j° at the same temperature. For the vapor-liquid equilibrium of the pure substance j we have to introduce the special values for the standard states, namely,

$$x_j = 1; \quad y_j = 1; \quad \gamma_j = 1; \quad \phi_j = \phi_j^o; \quad \overline{V}_j = \overline{V}_j^o; \quad P = p_j^o. \tag{7.4}$$

The fugacity coefficient ϕ_j^o of the pure vapor at the pressure p_j^o is different from unity if the vapor pressure is high. We obtain from (7.3) for the equilibrium between the pure liquid and the pure vapor of component j

$$\overline{G}_j^o + \int_1^{p_j^o} \overline{V}_j^o dP = \overline{G}_j^{o\prime} + RT\ell n(p_j^o \phi_j^o). \tag{7.5}$$

Subtracting this equation from (7.3) we get rid of $\overline{G}_j^o$ and $\overline{G}_j^{o\prime}$ and find

$$RT\ell n(x_j \gamma_j) + \int_1^P \overline{V}_j dP - \int_1^{p_j^o} \overline{V}_j^o dP = RT\ell n(Py_j\phi_j)/p_j^o\phi_j^o). \tag{7.6}$$

We introduce now the first simplification, permissible practically always except close to a critical point. In the Poynting integrals, which are minor correction terms, we replace $\overline{V}_j$ and $\overline{V}_j^o$ by the molal volume of the pure liquid component at atmospheric pressure. Thus we have instead of the integrals the term $\overline{V}_j^o(P - p_j^o)$ so that we can compute the activity coefficient

$$\ell n\, \gamma_j = \ell n(Py_j\phi_j)/p_j^o(x_j\phi_j^o) - \overline{V}_j^o(P - p_j^o)/(RT). \tag{7.7}$$

The second simplification is permissible if p_j^o has a moderate value, i.e., is in the range in which the compressibility factor (Fig. 3.3) is a linear function of the pressure. In this case Eq.(3.52) expresses the fugacity coefficient by means of the second virial coefficient β_j as

$$\ell n\, \phi_j^o = \beta_j^o p_j^o/(RT). \tag{7.8}$$

If in addition the total pressure P has a moderate value, and if the rule of Lewis and Randall (Section 6.21) may be assumed, the second virial coefficient is independent of the composition of the gaseous mixture and equal to its value for the pure component. Thus we obtain, as the result of the third simplification,

$$\ell n(\phi_j/\phi_j^o) = \beta_j^o(P - p_j^o)/(RT) \tag{7.9}$$

and from (7.7)

$$\ell n\, \gamma_j = \ell n(Py_j/p_j^o x_j) + (\beta_j^o - \overline{V}_j^o)(P - p_j^o)/(RT) . \tag{7.10}$$

The fourth simplification, often reasonable, is based on the assumption that the vapor is a perfect gas. As a rule, though not always, the liquid volume is much smaller than the absolute value of the second virial coefficient. For low pressures one may drop the last term in (7.10), so that

$$\gamma_j = Py_j/(p_j^o x_j) . \tag{7.11}$$

The very last simplification is based on the assumption that the liquid is a perfect solution; it leads to

$$y_j/x_j \quad = \quad p_j^o/P. \tag{7.12}$$

This relation furnishes sometimes a reasonable approximation for hydrocarbon mixtures. It may also be reasonable for a crude first orientation.

The separation factor, for vapor-liquid equilibria called relative volatility, is in the simplest case

$$\alpha_k^j \quad = \quad y_j x_k/(y_k x_j) \quad = \quad p_j^o/p_k^o \tag{7.13}$$

independent of composition and pressure; its dependence on the temperature is given, according to Clapeyron's Eq.(5.30), by the difference of the heats of vaporization. At a given temperature, the variation of the relative volatility is due only to the imperfection of the liquid and gaseous mixtures.

7.12 Two Examples. Observed data for vapor liquid equilibria are shown in Fig. 7.1 to 7.4. Both vapor composition and pressure of mixtures of benzene and cyclohexane [71.1] do not greatly deviate from the values of a perfect solution, indicated in the figures by broken lines; this is to be expected for a mixture of hydrocarbons. But the deviations are very large for ethanol-methylcyclohexane [71.2].

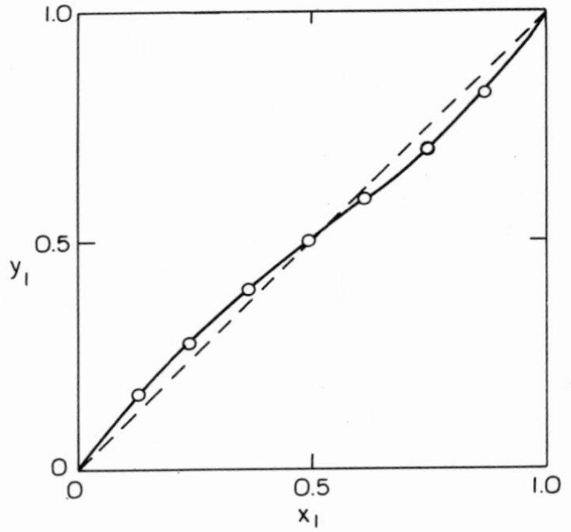

Fig. 7.1. Benzene-Cyclohexane (40.00 C. The broken line represents a perfect solution. G. Scatchard, S. E. Wood and J. M. Mochel, J. Phys. Chem. 43 (1939) 119)

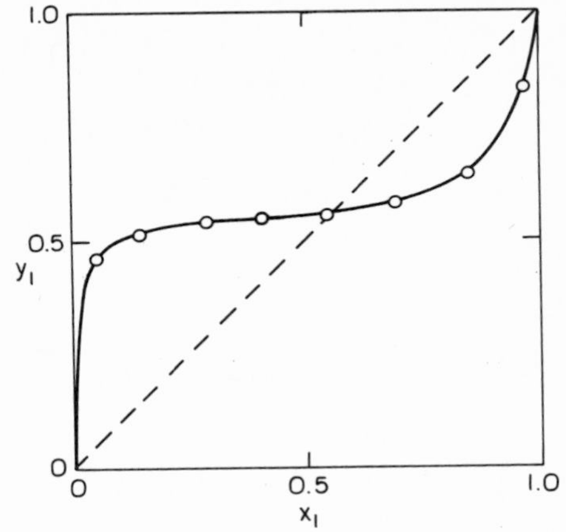

Fig. 7.2. Ethanol-Methylcyclohexane (35.00 C)

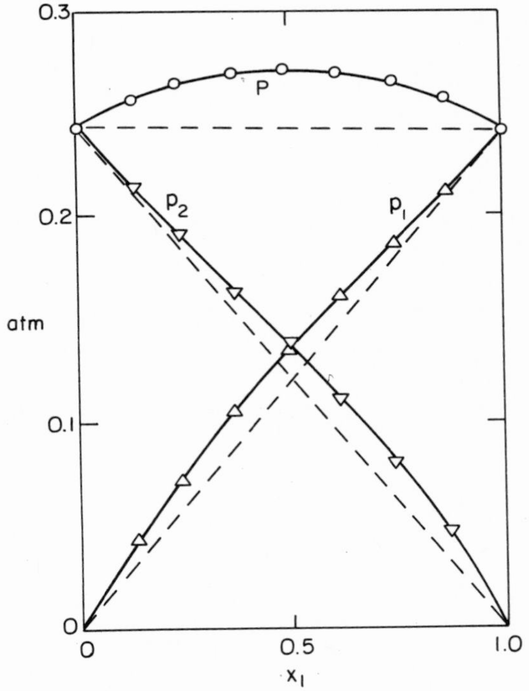

Fig. 7.3. Benzene-Cyclohexane (40.00 C. ○ Total pressure P; △▽ partial pressures p_1 and p_2. The broken lines represent a perfect solution)

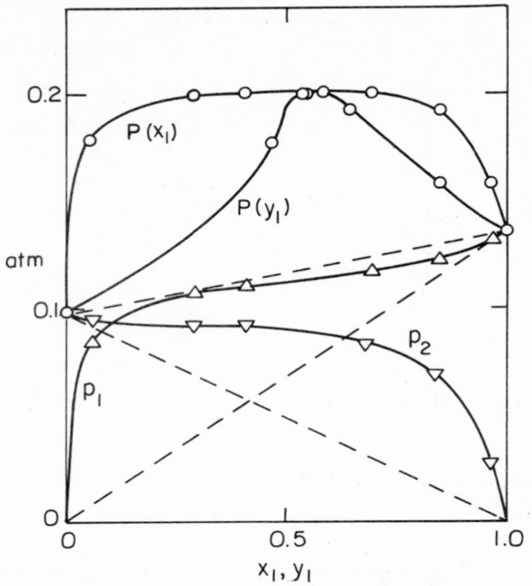

Fig. 7.4. Ethanol-Methylcyclohexane (35.00 C)

The computation of the relative volatility α and the activity coefficients γ_1 and γ_2 is outlined for three mixtures each in Tables 7.1 and 7.2.

Table 7.1

Vapor-Liquid Equilibrium Benzene(1)-Cyclohexane(2) at 40.00 C

(1)	P(atm)	0.24278	0.25650	0.27121	0.25714	0.24025
(2)	x_1	0	.1282	.4932	.8656	1
(3)	y_1	0	.1657	.4950	.8205	1
(4)	$\alpha = y_1x_2/(x_1y_2)$	---	1.3506	1.0072	.7097	---
(5)	$\log[y_1P/(x_1p_1^\circ)]$	---	.1398	.0542	.0063	0
(6)	VI1	-0.00006	-.00037	-.00070	-.00038	0
(7)	$\log \gamma_1$	(0.182)	.1395	.0535	.0059	0
(8)	$\log[y_2P/(x_2p_2^\circ)]$	0	.0048	.0465	.1507	---
(9)	VI2	0	-.00036	-.00073	-.00037	+0.00006
(10)	$\log \gamma_2$	0	.0044	.0458	.1503	(0.194)
(11)	$\log(\gamma_1/\gamma_2)$	(0.182)	.1351	.0077	-.1444	(-0.194)

VI1 = vapor imperfection correction of benzene = $(\beta_1^\circ - \bar{V}_1^\circ)/(2.303\ RT)$

VI2 = same for cyclohexane

Table 7.2

Vapor-Liquid Equilibrium Ethanol(1)-Methylcyclohexane(2) at 35.00 C

(1)	P(atm)	0.9687	0.17816	0.20047	0.15795	0.13571
(2)	x_1	0	.0526	.4052	.9676	1
(3)	y_1	0	.4645	.5471	.8369	1
(4)	$\alpha = y_1x_2/(x_1y_2)$	---	15.62	1.773	.1719	---
(5)	$\log[y_1P/(x_1p_1^o)]$	---	1.0645	0.2997	0.0030	0
(6)	VI1	+ .0005	- .0006	- .0009	- .0003	0
(7)	$\log \gamma_1$	---	1.0639	.2988	.0027	0
(8)	$\log[y_2P/(x_2p_2^o)]$	0	.0166	.1976	.9142	---
(9)	VI2	0	- .0028	- .0036	- .0021	- .0014
(10)	$\log \gamma_2$	---	.0138	.1940	.9121	---
(11)	$\log(\gamma_1/\gamma_2)$	---	+1.0501	+ .1048	- .9094	---

VI1 and VI2 = vapor imperfection corrections.

The first three lines contain the observed equilibrium data, including the vapor pressures p_2^o (first column) and p_1^o (last column) of the pure components. The 4th, 5th and 8th lines can be immediately computed from the data. One can see that the relative volatility varies moderately for the first set, but strongly for the second set.

The data for the corrections for vapor imperfection VI1 and VI2, as prescribed in Eq.(7.10) are given in Table 7.3. The second virial coefficients β_j^o have been estimated by D. Berthelot's relation (3.39); the numerical values are probably somewhat too low. The more elaborate estimates of the observers [71.1,71.2] are likely to be better. With the present estimates one obtains the corrections in lines 6 and 9 and the activity coefficients in lines 7 and 10. The correction terms are appreciable for the ethanol mixtures.

Table 7.3

Correction of Vapor Imperfection

	Benzene (40.00 C)	Cyclohexane (40.00 C)	Ethanol (35.00 C)	Methylcyclohexane (35.00 C)
T_c(K)	562.61	554.17	516.3	572.29
P_c(atm)	48.6	40.4	63.15	34.322
β^o(ml)	-1245	-1410	-745	-1895
$\bar{V}^o$(ml)	91.1	110.7	59.34	128.2
$(\beta^o-\bar{V}^o)/(2.303\ RT)$	-0.0226	-0.0255	-0.0138	-0.0348

The results for benzene-cyclohexane are shown in Figs. 7.5 and 7.6; those for ethanol-methylcyclohexane have been shown before in Figs. 6.3 and 6.4. The high accuracy of the data is indicated not only by the smoothness of the diagrams but also by consistency tests such as the area condition (6.41).

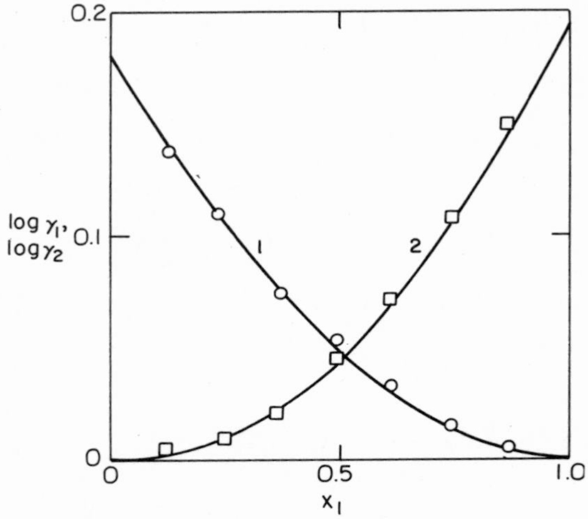

Fig. 7.5. Benzene-Cyclohexane (40.00 C; ○ log γ_1; □ log γ_2. Curves calculated with B = 0.188; C = 0.006)

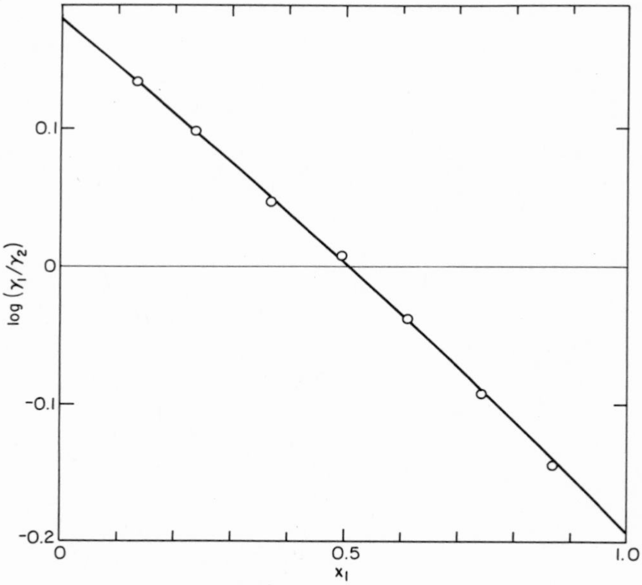

Fig. 7.6. Benzene-Cyclohexane

It can be seen from Figs. 7.5 and 7.6 that the activity coefficients of benzene-cyclohexane are excellently represented by a power series with two coefficients. As has been pointed out in Section 6.32, the results for ethanol-methylcyclohexane (Fig. 6.3) cannot even approximately be represented by a similar relation.

7.13 Partial Pressure; the Laws of Raoult and Henry. The term "partial pressure," though less often used today than in former times, is still convenient for gaseous mixtures at low presures. We define it as

$$p_j = Py_j \qquad (7.14)$$

and prefer to restrict its use to perfect gaseous mixtures, where this and other possible definitions coincide. But we do not rigorously observe this restriction.

In view of (6.13) we have for perfect mixtures

$$\bar{G}_j = \bar{G}_j^o + RT\ell n\ p_j \qquad (7.15)$$

so that p_j increases with the chemical potential $\bar{G}_j$, and conversely. The partial pressure can therefore substitute in qualitative questions for the chemical potential. Summation of (7.14) over all z components furnishes

$$p_1 + p_2 + \ldots + p_z = P. \qquad (7.16)$$

If not only the vapor but also the liquid solution is perfect, we conclude from (7.11) that

$$p_j = Py_j = p_j^o x_j, \qquad (7.17)$$

i.e., that the partial vapor pressure is proportional to the mole fraction in the liquid, with the vapor pressure p_j^o of the component as the proportionality factor. According to (7.16) the total pressure must then also be a linear function of the mole fractions. The corresponding straight lines are indicated by broken lines in Fig. 7.1 to 7.4. The deviations are moderately small for benzene-cyclohexane, they are very large for ethanol-methylcyclohexane.

For a perfect vapor and imperfect liquid we write (7.11) as

$$p_k = p_k^o x_k \gamma_k \qquad (7.18)$$

and differentiate with respect to x_j. The result

$$dp_k/dx_k = p_k^o \gamma_k + p_k^o x_k \cdot d\gamma_k/dx_k \qquad (7.19)$$

is interesting at the limits $x_k = 1$ and $x_k = 0$.

For the pure component $k(x_k = 1)$ we have by definition (Section 6.31)

$$\gamma_k = 1 \tag{7.20}$$

and from the relation of Gibbs (Eq.(6.34)) we conclude

$$d\gamma_k/dx_k = 0$$

since all x_j's with exception of x_k must be zero for $x_k = 1$. We conclude therefore from (7.19) that

$$\lim_{x_k=1} dp_k/dx_k = p_k^o. \tag{7.21}$$

Accordingly, in Figs. 7.3 and 7.4 the curves for p_1 and p_2 are tangent to the broken lines at $x_1 = 1$ and $x_2 = 1$, respectively. In other words, the partial vapor pressure of the solvent of <u>any</u> dilute solution (x_k close to 1) behaves like that of the solvent of a perfect solution. This is <u>Raoult's Law</u>.

Its general validity as a limiting law was important in early chemistry. Indeed, molal weight determinations were based either on Raoult's law or on another limiting law, that of the perfect gas.

For a dilute solution of component k we conclude from (7.19)

$$\lim_{x_k=0} dp_k/dx_k = p_k^o \lim_{x_k=0} \gamma_k = h \tag{7.22}$$

so that we obtain as an approximation

$$p_k = hx_k . \tag{7.23}$$

This is <u>Henry's Law</u>. It differs from Raoult's law in that the coefficient h is an empirical factor different from p_k^o (if the solution is not perfect). Henry's law means that the curve for p_1 in Figs. 7.3 or 7.4 is approximately represented for small values of x_1 by the limiting tangent of p_1 for $x_1 = 0$. This approximation is obviously useful in a case such as Fig. 7.3, but of less value for a highly imperfect solution as shown in Fig. 7.4.

7.14 <u>The Total Pressure.</u> The investigation of vapor-liquid equilibria comprises usually the measurements of the composition of both phases in addition to the temperature and pressure. It has been shown in Tables 7.1 and 7.2 how the activity coefficients of a binary solution are derived from those data. The existence of a thermodynamic condition, namely, the relation of Gibbs, Duhem and Margules, shows that the experimental data are redundant. Practically one wishes, as a rule, to drop the determination of the vapor composition and to derive the activity coefficients from measurements of the total pressure as a function of the composition of the liquid at a given temperature.

We restrict the discussion to the case of a perfect vapor and start from Eq.(7.11) which we write now for a binary solution as

$$Py_1 = x_1\gamma_1 p_1^\circ; \qquad Py_2 = x_2\gamma_2 p_2^\circ \tag{7.24}$$

so that

$$P = x_1\gamma_1 p_1^\circ + x_2\gamma_2 p_2^\circ \ . \tag{7.25}$$

This is a relation between the observed quantities x_1, x_2, P, p_1°, p_2°, and the unknown activity coefficients γ_1 and γ_2. One can solve for both unknowns with the aid of the equation of Gibbs (6.34)

$$x_1 \cdot d\ell n\ \gamma_1/dx_1 + x_2\ d\ell n\ \gamma_2/dx_1 = 0 \ . \tag{7.26}$$

The problem has been early recognized [71.3] and often discussed [71.4]. Essentially there are two ways of solution.

One can choose an algebraic representation such as (6.50) and (6.51) for γ_1 and γ_2 and determine the values of the coefficients from (7.25) and (7.26) by trial and error.

Alternatively, one derives from (7.24) at constant temperature

$$\frac{d\ell n\ \gamma_1}{dx_1} = \frac{d\ell n\ P}{dx_1} + \frac{d\ell n\ y_1}{dx_1} - \frac{1}{x_1}\ ; \qquad \frac{d\ell n\ \gamma_2}{dx_1} = \frac{d\ell n\ P}{dx_1} + \frac{d\ell n\ y_2}{dx_1} - \frac{1}{x_2} \tag{7.27}$$

and introduces these derivatives into (7.26). The result is the relation of Lewis and Murphree

$$dy_1/dx_1 = d\ell n\ P/dx_1 \cdot y_1 y_2/(y_1 - x_1). \tag{7.28}$$

The integration of this equation furnishes y_1 as function of x_1, i.e., the basis for conventional computation of γ_1 and γ_2.

The application of both methods is easy for nearly perfect solutions, but it can be difficult [71.5] for solutions that widely deviate from perfection.

The relationship between the two phases for a binary solution is most clearly seen in a diagram such as shown in Fig. 7.7. If we start from the initial point I, representing a mixture of a given composition under a high pressure, and decrease the pressure isothermally, we arrive at the <u>bubble point</u> B and vapor of the equilibrium composition D develops. The mole fraction x_1 of the liquid decreases due to the loss of the vapor and the liquid is represented on further pressure decrease along the <u>bubble line</u> B-L. At the same

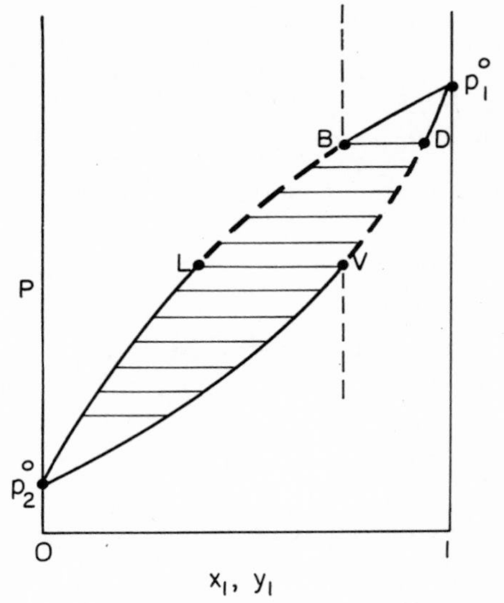

Fig. 7.7. Isothermal Vaporization

time more vapor is formed and the composition of the vapor follows the dew line D-V. The phases in equilibrium are connected by the horizontal tielines. When the vapor reaches the original composition of the mixture at V no liquid is left and further pressure decrease leads into the field of the unsaturated gaseous mixture.

In order to find the bubble pressure P_B for a liquid of mole fraction x_1 we add the two equations (7.24) and obtain

$$P_B = x_1\gamma_1 p_1^o + x_2\gamma_2 p_2^o . \qquad (7.29)$$

Similarly we can find the dew pressure P_D for a vapor composition y_1. From (7.24) we conclude that

$$x_1/P_D = y_1(\gamma_1 p_1^o); \qquad x_2/P_D = y_2(\gamma_2 p_2^o) \qquad (7.30)$$

so that

$$1/P_D = y_1/(\gamma_1 p_1^o) + y_2/(\gamma_2 p_2^o). \qquad (7.31)$$

For multicomponent mixtures one derives in a similar manner from (7.11)

$$P_B = \Sigma(x_j\gamma_j p_j^o); \qquad 1/P_D = \Sigma y_j/(\gamma_j p_j^o). \qquad (7.32)$$

Information of the activity coefficients is presupposed for the application of these relations. The dew point calculation requires a step by step approximation.

7.15 Azeotropic Solutions. In both examples for vapor-liquid equilibrium (Figs. 7.1 and 7.2) there is one mixture (x_1 = 0.53 and 0.55, respectively) for which

$$x_1 = y_1 \qquad (7.33)$$

i.e., both phases have the same composition. Such mixtures are called azeotropic or constant boiling mixtures since the boiling temperature under a given pressure (or the boiling pressure at a given temperature) cannot change in the progress of a distillation. According to the phase rule only two intensive variables are independent in a two-phase equilibrium of two components; as long as composition and pressure do not change, no other intensive quantity can.

For a binary azeotrope with perfect vapor we obtain from (7.11)

$$\gamma_1 = P/p_1^o; \qquad \gamma_2 = P/p_2^o; \qquad \gamma_1/\gamma_2 = p_2^o/p_1^o; \qquad \alpha = 1. \qquad (7.34)$$

In a way, the existence of an azeotropic point is accidental since γ_1/γ_2 and p_2^o/p_1^o express properties entirely unrelated to each other. But azeotropes are practically important because their components cannot be separated by simple distillation.

The relations (7.34) lead to a convenient survey of the appearance of azeotropic mixtures in Figs. 7.8 and 7.9. The relation between log α and $\log(\gamma_1/\gamma_2)$ is illustrated in Fig. 7.8. If the vapor is perfect the two lines are parallel with the constant vertical distance $\log(p_1^o/p_2^o)$. The log α curve crosses the zero line if $\log(p_1^o/p_2^o)$ is equal to $-\log(\gamma_1/\gamma_2)$ for some value of x_1. This happens if $\log(p_1^o/p_2^o)$ is between the limits $-L'$ and L'' of $\log(\gamma_1/\gamma_2)$. The vapor pressure diagram 7.9 shows that the temperature range for an azeotrope lies between T' and T"; with increasing temperature, an azeotrope appears at T' and $x_1 = 0$, its mole fraction increases, and at T" it disappears at $x_1 = 1$.

Actually the heats of vaporization of substances with close vapor pressures are, as a rule, not much different; therefore the angle between the curves is small and an azeotrope exists in a large temperature range. But this is not always so.

An azeotrope must exist if the vapor pressure curves cross. This, however, is not a necessary condition. Propanol-2 and water form an azeotrope though propanol-2 has a higher vapor pressure at all temperatures [71.6].

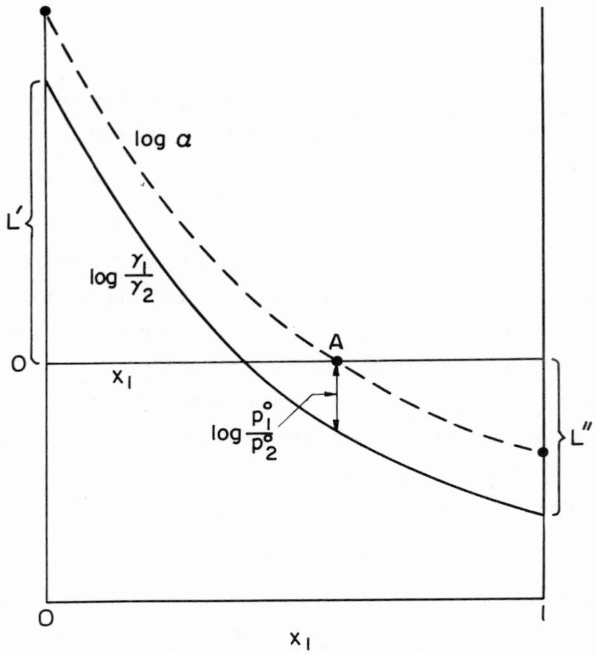

Fig. 7.8. Azeotrope A

A perfect solution (L' = 0, L" = 0) is azeotropic only at the crossing point of the vapor pressure curves. Large deviations from the perfect solution result in a wide range of azeotropy.

In the examples of Figs. 7.3 and 7.4 the total pressure is at a maximum at the azeotropic point. It can indeed be shown that the pressure at an azeotropic point must have either a maximum or a minimum value (Konovalov's rule). From (7.25) we obtain by differentiation at constant temperature

$$dP/dx_1 = \gamma_1 p_1^o - \gamma_2 p_2^o + x_1 p_1^o \cdot d\gamma_1/dx_1 + x_2 p_2^o \cdot d\gamma_2/dx_1. \tag{7.35}$$

For an azeotropic point we can eliminate p_1^o and p_2^o by means of (7.34). The result is in view of (7.26)

$$(dP/dx_1)_{az} = 0. \tag{7.36}$$

One concludes from (7.25) and Fig. 7.3 that the extremal value is a maximum if the activity coefficients are greater than unity, the usual case for non-electrolytes. Activity coefficients below unity happen rarely for non-electrolyte mixtures; they indicate a compound formation between the components.

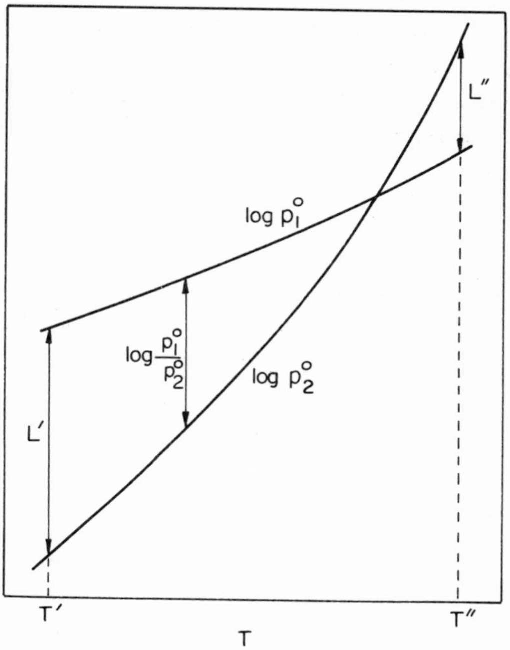

Fig. 7.9. Azeotropic Temperature Range [there is an azeotrope as long as L' > $\log(p_1^o/p_2^o)$ and L" > $\log(p_2^o/p_1^o)$]

Crossplotting a set of isothermal pressure curves furnishes isobars that show a minimum temperature for a maximum pressure and conversely.

Since azeotropic solutions cannot be separated by simple distillation one has to use other methods such as extraction. Sometimes the separation can be achieved by distillation with an auxiliary substance chosen so that the ratio γ_1/γ_2 is efficiently changed ("azeotropic distillation" if the third component is volatile at the given conditions, "extractive distillation" if it is non-volatile).

7.2 LIQUID-LIQUID EQUILIBRIUM

7.21 Incompletely Miscible Systems. According to the phase rule (Section 5.2) a system of two components in three phases has still one degree of freedom. In other words, if two components are only partially soluble in each other and vapor is present, all intensive properties, including the pressure, are fixed by the choice of the temperature. Fig. 7.10 shows the mutual solubility of liquid sulfur dioxide and octane as a function of the temperature. In the area beneath the solubility curve the mixture splits into two phases; their compositions are given by the curve. A tie line is shown as an example at 0 C.

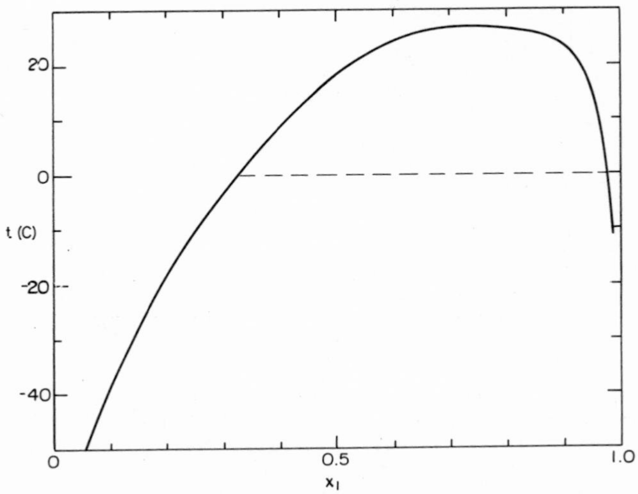

Fig. 7.10. Mutual Solubility of Sulfur Dioxide and Octane (Upper critical solution point. Data: W. F. Seyer and A. F. Gallagher, Trans. Roy. Soc. Canada, 20 (1926) 343)

With increasing temperature the tie lines become shorter, i.e., the two saturated solutions become more similar. At 27 C the solutions conjugated by the tie line are identical in a critical solution point near $x_1 = 0.73$. Above this temperature only a single liquid phase exists.

The condition for equilibrium between the two liquid phases is given by (7.1). If we denote the two phases by single and double primes we have

$$\bar{G}_1' = \bar{G}_1''; \qquad \bar{G}_2' = \bar{G}_2''. \tag{7.37}$$

This condition may be converted to a relation between activity coefficients by means of (6.29). The standard state of the first component is the same for both phases, namely, the pure liquid. Similarly, there is only one standard state for the second or any other component. The Poynting integrals usually may be disregarded, especially since both phases are under the same pressure. The result of introducing (6.29) into (7.37) is then that the activities

$$x_j'\gamma_j' = x_j''\gamma_j'' \tag{7.38}$$

of each component in both phases are equal. For a binary system this equilibrium condition can be used to determine two parameters in an algebraic representation, e.g., B and C in (6.50) and (6.51), if such a representation can be justified, i.e., as a rule for non-associating substances of low or moderate polarity.

The characteristic features of the activity coefficients of immiscible solutions can be recognized even in the simplest case of a one-parameter Eq.(6.47), although such a relation will rarely be adequate for a representation of such highly imperfect mixtures. The diagram of Fig. 7.11 shows

$$\log(x_1\gamma_1) = \log x_1 + Bx_2^2 \tag{7.39}$$

for the values B = 0.5, 0.7, 0.868, 1.0. The two lower curves are monotonically increasing: there is no pair of mole fractions for which $\log(x_1\gamma_1)$ would assume the same value. Such pairs, however, are found in the highest curve. For B = 1.0, we find, for instance, the activity -0.052 for x_1'=0.215

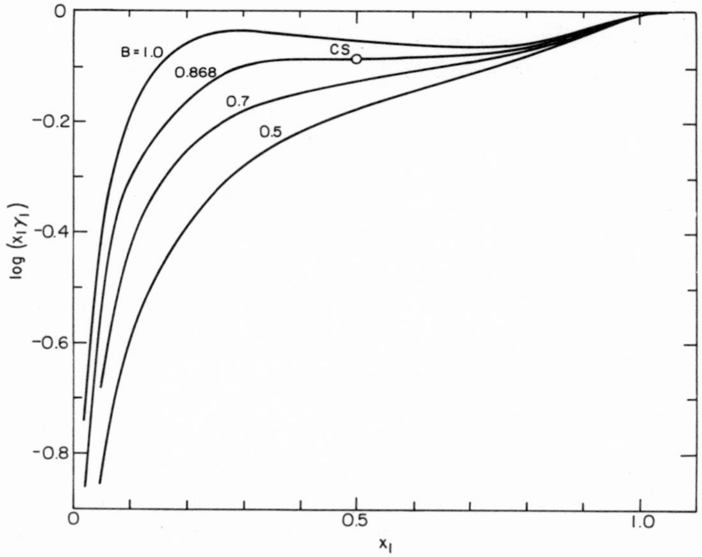

Fig. 7.11. Activity in the Vicinity of a Critical Solution Point CS

as well as for $x_1'' = 0.830$. The border line between the two types of curves is represented by the curve for B = 0.868; this curve shows a horizontal inflection tangent at $x_1 = 0.5$. Whenever this happens the mixture is a border case between completely and incompletely miscible solutions. This point of the horizontal inflection tangent is a <u>critical solution point</u>. Qualitatively, though not quantitatively, Fig. 7.11 corresponds with Fig. 7.10, the critical isotherm at 27 C would be the curve for B = 0.868. The curves with lower values of B indicate the behavior of the activity at higher temperatures, the curve for B = 1.0 at some temperature lower than 27 C.

7.22 <u>Stability</u>. The similarity of Fig. 7.11 and Fig. 3.3 is not accidental: Both describe the equilibrium between two phases (vapor-liquid or liquid-liquid). The equilibrium condition is the equality of the molal (or partial molal) free energies. In the vapor-liquid equilibrium it is the pressure that represents the molal free energy, in the liquid-liquid equilibrium it is the activity; both functions are monotonically increasing with the free energy. In the families of curves of both diagrams the opportunity of equal values of the (partial) molal free energy for two different coordinate values disappears at the critical isotherm. Since the critical point is the limit of two coexisting phases ($\bar{G}_j' = \bar{G}_j''$) the derivative of $\bar{G}_j$ with respect to the generalized coordinate must be zero. This implies

$$\partial(x_1\gamma_1)/\partial x_1 = 0 \quad \text{(crit.)} \tag{7.40}$$

so that the curves are horizontal in the critical solution point.

Moreover, the critical curve (B = 0.868) in Fig. 7.11 cannot bend down above the critical composition because in this case we should again have equal activities in two different points. For this reason the critical tangent must be an inflection tangent

$$\partial^2(x_1\gamma_1)/\partial x_1^2 = 0 \tag{7.41}$$

which is the same condition as (3.28).

Since the bending down of a curve fo $x_1\gamma_1$ creates the opportunity for equal activities in two phases, the stability of a phase with respect to a split into two phases is ensured by

$$\partial(x_1\gamma_1)/\partial x_1 > 0 \, . \tag{7.42}$$

This condition can be easily illustrated (Fig. 7.12) by introducing (7.18) into (7.42); a perfect vapor is assumed for simplification. In the first case, satisfying (7.42), component 1 distills from the second solution, which

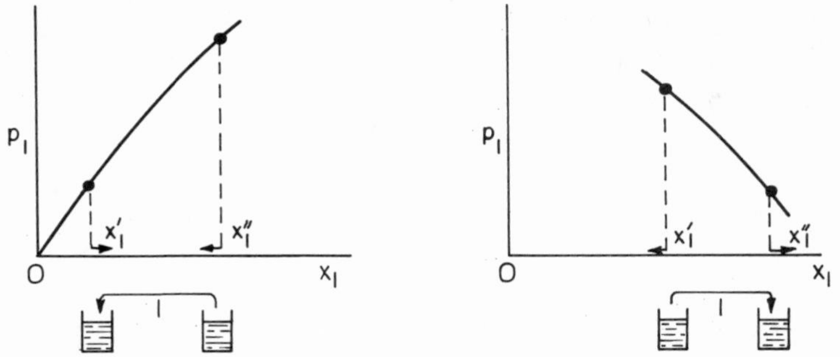

Fig. 7.12. A Stable and an Unstable Vapor Pressure Curve

has the higher partial vapor pressure, to the first so that x_1'' decreases and x_1' increases; the final, stable result is a single solution with a mole fraction between x_1' and x_1''. However, if the opposite of (7.42) happens, component 1 distills from the first to the second solution so that the more concentrated solution becomes still more concentrated and the concentration gap increases. Equilibrium between two liquid phases is reached only when the partial vapor pressures have become equal. The same considerations hold for any generalized force.

For these reasons we call stability condition the requirement that the derivative of a generalized force with respect to its conjugate coordinate must have a positive value.

Around a critical solution point, Eq.(7.42) requires, in view of (7.40), that $\partial^2(x_1\gamma_1)/\partial x_1^2$ changes its signs. Condition (7.41), however, is stronger than necessary; there may be a jump from negative to positive values.

The critical solution temperature is usually the upper limit of the range of equilibrium between two phases. Much less frequent is a lower critical solution point such as shown in Fig. 7.13.

7.23 Extraction. Liquid-liquid equilibria are used for the separation of two components, particularly when distillation is difficult or inapplicable (azeotropic mixture or low vapor pressure). For separation by extraction one adds a solvent (S in Fig. 7.14) that is not completely miscible with at least one component (A) and much more or completely miscible with the other (B in the diagram). The proportions of the components which lead to two liquid phases are indicated by the field enclosed by the saturation line R, R_3, R_2, R_1, P, E_1, E_2, E_3, E. If 1 mole feed is mixed with 2 moles solvent the diagram indicates the mixture at M_1. This mixture splits in equilibrium into the raffinate R_1, containing less B than the feed, and the extract E_1. Mixing of raffinate with more solvent furnishes M_2, which splits in R_2 and E_2. By repetition one can approach R.

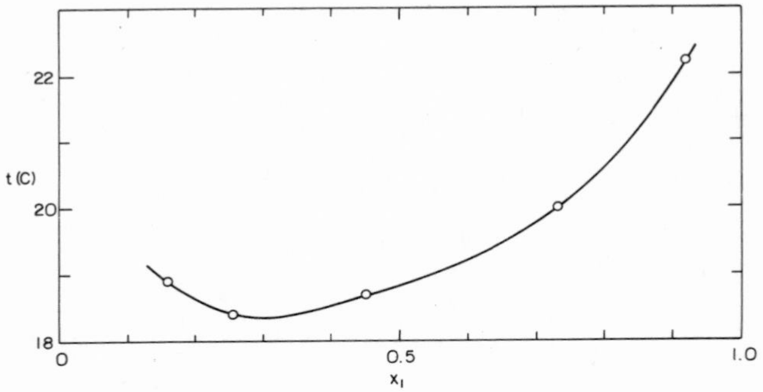

Fig. 7.13. Mutual Solubility of Triethylamine and Water (Lower critical solution point. Data: L. D. Roberts and J. E. Mayer, J. Chem. Phys. 9 (1941) 852)

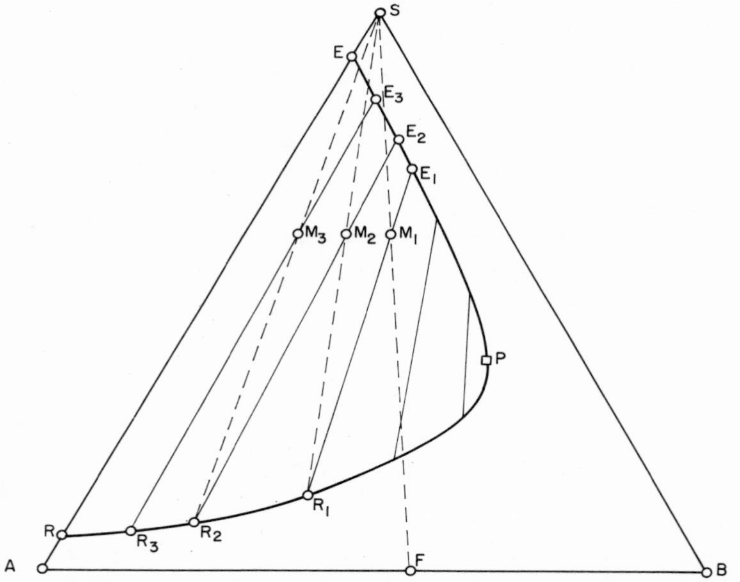

Fig. 7.14. Extraction (A, B components to be separated, S solvent, F feed; M_1, M_2, M_3 mixtures with solvent; E_1, E_2, E_3 extracts; R_1, R_2, R_3 raffinates; R_1 - E_1 etc. tie lines; P plait point)

An important criterion of the efficiency of extraction in a given case is the separation factor ('raffinate, "extract)

$$\alpha_2^1 = x_1''x_2'/(x_1'x_2'') = \gamma_1'\gamma_2''/(\gamma_1''/\gamma_2'). \tag{7.43}$$

A first orientation of the activity coefficients can often be obtained by appropriate superposition of the functions Q (Eq.(6.44)) for the three binary border systems. But adequate design, as a rule, will require also some experimental study.

In the <u>plait point</u> P (Fig. 7.14) the tie line shrinks to a point. This point has the character of a critical point as the border between a single phase region and two phases in equilibrium approaching identity. The separation factor α_2^1 goes to unity at the plait point.

7.3 THE CRITICAL REGION

If the temperature is raised beyond the critical temperature of the lower boiling (conventionally the first) component, the diagram of Fig. 7.7 does not apply any more. Since no pure liquid 1 exists the vapor pressure diagram for a given temperature shows the pattern of Fig. 7.15. Obviously the mixture has now a critical point C, which is the limit of the tie lines and the border between the two-phase region and the gaseous phase.

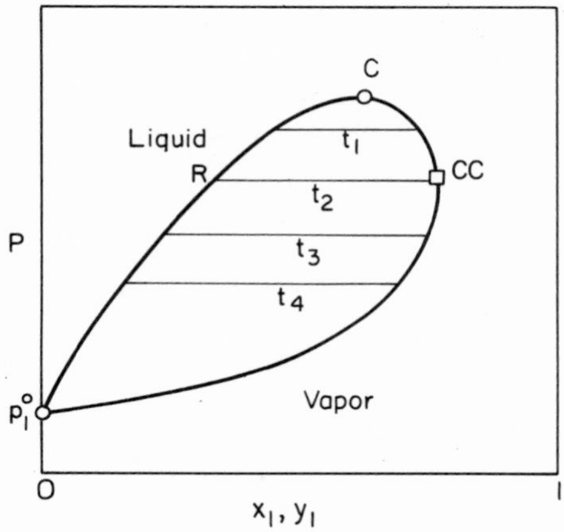

Fig. 7.15. Vapor Pressure (Temperature between the critical temperatures of the components. C critical point, CC critical condensation point, t_1, t_2 ... tie lines)

The reduction of pressure from an initial high-pressure point proceeds in the same manner as in Fig. 7.7 as long as we start at a mole fraction lower than that of R. We encounter a bubble line at the left of R and a dew line below the point CC.

A second phase separates out also if we start from a liquid in the composition range R-C. Contrary to the case of Fig. 7.7, the second phase becomes, on continued pressure reduction, more concentrated in the first component until it reaches the point CC.

If the composition of the mixture is in the range between C and CC, the reduction of the initial high pressure leads to a point on the line C-CC. Here a liquid, lower in the first component separates out. The separation of a liquid on reduction of the pressure is called retrograde condensation (the same happens in this case on isobaric temperature increase). Vaporization starts when the pressure is reduced below the value at point CC and progresses at further reduction.

The lower limit CC of the region of retrograde condensation has been called the critical condensation point though this name has not been generally adopted. At this point the highest concentration of component 1 in the vapor of a two-phase equilibrium is reached at the given temperature. From Fig. 7.15 one concludes immediately that at constant temperature

$$(dP/dy_1)_{CC} = \infty . \qquad (7.44)$$

The upper limit C of the retrograde condensation range is the critical point. It obviously satisfies the condition

$$(dP/dy_1)_C = 0 \qquad (7.45)$$

at constant temperature. The same argument that leads to (7.40) and (7.41) also results in

$$\partial(y_1\phi_1)/\partial y_1 = 0; \qquad \partial^2(y_1\phi_1)/\partial y_1^2 = 0 \qquad (7.46)$$

for the critical point of a solution. These conditions are sufficient for deriving from an equation of state the critical temperature and pressure as functions of the composition [73.1].

Retrograde condensation and critical phenomena of solutions in general have been of technical interest because oil production has progressed to deep wells where temperature and pressure often are high enough to produce critical conditions.

A general aspect of the liquid-vapor equilibrium is given in Fig. 7.16.

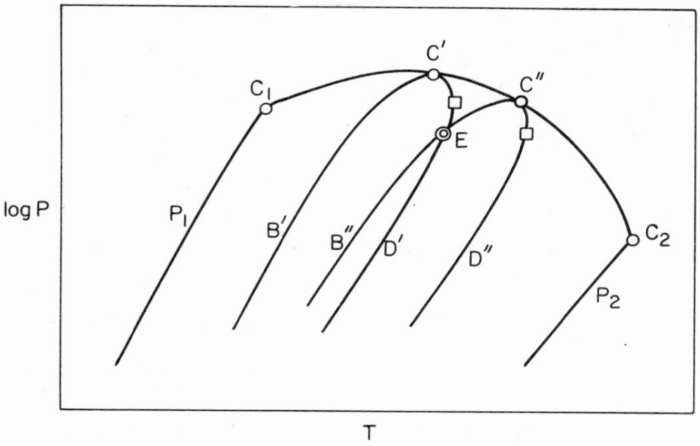

Fig. 7.16. Critical line $C_1C'C''C_2$ (P_1, P_2 vapor pressure curves of the pure components; B', B" bubble curves; D', D" dew curves. The liquid of B" and the vapor of D' have equal temperature and pressure at E, i.e., there they are in equilibrium)

The field in which two phases may coexist is limited by the vapor pressure curves P_1 and P_2 of the components, and the critical line $C_1C'C''C_2$. For a given mole fraction x' or y' the bubble and dew lines are indicated by B'C'D'; they mean that a liquid of mole fraction x_1' is stable at pressures above B', and that a gas of mole fraction y_1' is stable at pressures below D'. The point E marks the equilibrium between a liquid x" and a vapor y' at the given temperature and pressure. Since bubble lines as well as dew lines completely fill the two-phase field, every point marks the equilibrium of a liquid and a vapor. (If the mole fraction x_1 or y_1 is plotted perpendicularly to the diagram of Fig. 7.16, each point in the diagram is the projection of a tie line).

The critical point is the limit of the equilibrium point E when the concentration difference between the two phases disappears, i.e., for the limit $y_1'' = x_1'$ ($= x_1'' = y_1'$). The bubble-dew line is tangent to the critical line but the tangent is (in general) not horizontal. The critical condensation point (marked by a square in Fig. 7.16), however, has a vertical tangent also in this diagram because condition (7.44) implies $(\partial P/\partial T)_{y_1} = \infty$.

The pattern is more complicated if the system is azeotropic up to the critical region [73.2].

The critical point of a solution differs from that of a pure substance in that it marks the limit of stability with respect to <u>composition</u>. In this regard it is similar to the critical solution point of two liquid phases.

7.4 FREEZING POINT DEPRESSION

The equilibrium between a liquid solution and a solid was the basis of early operations in chemical technology as well as in chemical analysis and physical chemistry. Interest was drawn first to dilute solutions. As a rule, we still talk of the solvent (conventionally the first component), and of the solute (second component). The depression of the freezing point is the subject of a discussion of the equilibrium of the liquid solution with the solid solvent, and solubility is the subject of the equilibrium with the solid solute.

The basic thermodynamic relation is the same for both subjects, namely,

$$\bar{G}_j - \bar{G}_j' = 0 \tag{7.47}$$

as prescribed by (7.1). Solid solutions (mixed crystals) are of considerable interest in metallurgy and a few other problems such as urea-paraffin complexes, but they are not often quantitatively discussed. In the following we restrict ourselves to a pure solid phase and a given low (e.g., atmospheric) pressure so that Poynting terms may be disregarded. Thus the solid phase is assumed to be always in its standard state and its Gibbs free energy is $\bar{G}_j^{o\prime}$.

The essential thermodynamic relations do not know any distinction of solvent and solute. The distinction, however, is introduced into our equations as soon as we abandon for the solute the convention (Section 6.31) that the activity coefficient is unity for the pure substance. If the interest is concentrated on dilute solutions, particularly in water, we introduce for historical and practical reasons the molality (number of moles of solute per 1000 g solvent), given for non-electrolytes by

$$m = 1000x_2/(\bar{M}_1 x_1); \qquad \text{for aqueous solutions: } m = 55.506x_2/x_1. \tag{7.48}$$

The molal weight $\bar{M}_1$ is expressed in gram/mole. At the same time we usually choose for the activity a_2 and the activity coefficient

$$a_2 = m\gamma_2; \qquad \lim_{x_1=1} a_2 = m; \qquad \lim_{x_1=1} \gamma_2 = 1 \tag{7.49}$$

in contrast to the earlier stipulation which prescribed unity for a_2 and γ_2 at $x_2 = 1$. The general relation

$$\bar{G}_2 = \bar{G}_2^\circ + RT\ln a_2 \qquad (7.50)$$

now defines $\bar{G}_2^\circ$ as the molal chemical potential in a solution of unit activity, i.e., somewhere near m = 1.

7.41 Solvent Activity. The calculation of the activity of the solvent from observed freezing temperatures may be considered to be the central problem. According to the phase rule, equilibrium between the liquid and the solid phase in a binary system allows two independently variable intensive properties. Since we fix the pressure at one atmosphere, there is only one independent intensive property. For any mole fraction x_1 there exists an equilibrium freezing temperature T_e, which can be measured.

The activity of the solvent can be found from the equilibrium

$$\text{Solvent (s)} = \text{Solvent } (x_1 \text{ in liquid solution}). \quad \Delta G = 0. \qquad (7.51)$$

It will be important in the following to watch for the superscripts ° (standard state: pure solvent) and ' (solid state).

The equilibrium condition is

$$\Delta G = \bar{G}_1 - \bar{G}_1^{\circ\prime} = \Delta G^\circ + RT\ln a_1 = 0 \qquad (7.52)$$

with

$$\bar{G}_1 = \bar{G}_1^\circ + RT\ln a_1; \qquad \bar{G}_1^\circ - \bar{G}_1^{\circ\prime} = \Delta G^\circ. \qquad (7.53)$$

Here ΔG° is zero for equilibrium of the two pure phases, i.e., at the melting temperature T_f or temperature of fusion. The properties $\bar{G}_1^\circ$ and $\bar{G}_1^{\circ\prime}$ of the pure phases depend only on the temperature, but a_1 is a function of both T and x_1.

From (7.53) we derive by partial differentiation with respect to T, in view of (3.8),

$$\left(\frac{\partial(\bar{G}_1/T)}{\partial T}\right)_{x_1} = \frac{d(\bar{G}_1^\circ/T)}{dT} + R\left(\frac{\partial \ln a_1}{\partial T}\right)_{x_1}$$

$$= -\frac{\bar{H}_1}{T^2} = -\frac{\bar{H}_1^\circ}{T^2} + R\left(\frac{\partial \ln a_1}{\partial T}\right)_{x_1} . \qquad (7.58)$$

We call

$$\bar{L}_1 = \bar{H}_1 - \bar{H}_1^\circ \qquad (7.59)$$

the relative partial molal heat content: it is the increase in heat content on addition of 1 mole solvent to a large amount of solution of mole fraction x_1. The temperature coefficient of the activity is given by

$$\left(\frac{\partial \ln a_1}{\partial T}\right)_{x_1} = -\frac{\bar{L}_1}{RT^2} . \qquad (7.60)$$

But the differentiation of relations (7.52) or (7.53) can be carried out also along the equilibrium, i.e., while the condition $\Delta G = 0$ is maintained and x_1 is varying with T as implicitly prescribed by this condition. From (7.52) we derive

$$\left(\frac{d\ell n\ a_1}{dT}\right)_{eq} = -\frac{1}{R}\cdot\frac{d(\Delta G^o/T)}{dT} = \frac{\Delta H^o}{RT^2} \quad (7.61)$$

with the heat of fusion of the pure solvent

$$\Delta H^o = \bar{H}_1^o - \bar{H}_1^{o\prime} . \quad (7.62)$$

Integration of (7.61) furnishes a_1 as a function of x_1 at the equilibrium temperature T_e

$$\ell n\ a_1(T_e) = \int_{T_f}^{T_e} \frac{\Delta H^o}{RT^2}\, dT = -\int_{T_e}^{T_f} \frac{\Delta H^o}{RT^2}\, dT \quad (7.63)$$

as soon as the freezing point T_e is measured as a function of x_1. The activity is unity for the pure solvent; in this case the equilibrium temperature is T_f.

For any temperature T we obtain the activity from (7.60) as

$$\ell n\ a_1(T) = \ell n\ a_1(T_e) - \int_{T_e}^{T} \frac{\bar{L}_1}{RT^2}\, dT$$

$$= -\int_{T_e}^{T_f} \frac{\Delta H^o}{RT^2}\, dT - \int_{T_e}^{T} \frac{\bar{L}_1}{RT^2}\, dT. \quad (7.64)$$

For the fusion temperature T_f we obtain, in view of (7.59),

$$\ell n\ a_1(T_f) = \int_{T_e}^{T_f} \frac{\Delta H}{RT^2}\, dT \quad (7.65)$$

writing

$$\Delta H = \bar{H}_1 - \bar{H}_1^{o\prime} = \Delta H^o + \bar{L}_1 \quad (7.66)$$

for the heat of fusion of the solvent in a large amount of solution.

The application of these relations at two levels of accuracy will be discussed in the following.

7.42 A Metallurgical Example. A large body of data for freezing curves of binary alloys [74.1] furnishes information of their molecular constitution. Some of them can be used for the determination of the activity. The limited accuracy of the observations allows far reaching simplifications in the calculation.

For the system bismuth-cadmium we find the data given in Table 7.4;

Table 7.4

Bismuth(1)-Cadmium(2)

x_1	1 Bi	0.7	0.546 Eutectic		0.3	0 Cd
Freezing Temp(K)	544.4	476	419		479	594
ΔH°(cal)	2600					1460
$1000(1/T_e - 1/T_f)$		.265	.551	.703	.404	
log a_1	0	-.151	-.313			
γ_1	1	1.01	1.07			
log a_2				-.224	-.129	0
γ_2				1.09	1.06	1

the equilibrium temperatures of $x_1 = 0.3$ and $x_1 = 0.7$ have been taken from a graph. The two metals do not form solid solutions. The nearly linear freezing curves from the pure metals to the eutectic at $x_1 = 0.546$; 419 K indicate that no complications should be expected. We assume that the heats of mixing, and therefore $\bar{L}_1$ and $\bar{L}_2$ are negligible, and that the heats of fusion at the melting temperatures can be used over the whole range of calculation. The activities are therefore implicitly assumed to be independent of the temperature.

With these simplifications, (7.64) furnishes

$$\log a_1 \quad = \quad \frac{\Delta H^\circ}{2.303\ R}\left(\frac{1}{T_e} - \frac{1}{T_f}\right) \tag{7.67}$$

between bismuth and the eutectic, and the same relation (with different numerical values of ΔH° and T_f) for log a_2 between cadmium and the eutectic. If one wishes to find the activities or activity coefficients of both components over the whole range, one can make use of the relation (6.34) of Gibbs and Duhem. In general it is preferable to determine the parameters of a suitable algebraic representation, such as (6.50) and (6.51); this furnishes then complete information. In both cases the computation should start from a larger number of points, preferably from all immediate observations.

7.43 Aqueous Solutions. Freezing point depressions of dilute aqueous solutions have been extensively measured in the investigations of electrolytes. The general thermodynamic discussion has been developed by Lewis and

Randall. In this Section, still restricted to non-electrolytes, the activity coefficients of water and urea in their solutions will be derived from excellent measurements of the freezing points of urea [74.2].

The first step is the computation of the activity a_1 of water at the freezing temperature T_e. We introduce the numerical values [74.3] for the heat of fusion at 273.15 K

$$\Delta H^\circ(T_f) \;=\; 1436 \text{ cal/mole}, \tag{7.68}$$

for the change in heat capacity on fusion

$$\Delta C_p^\circ \;=\; 9.1 \text{ cal/deg·mole} \tag{7.69}$$

and accordingly

$$\Delta H^\circ \;=\; 1436 + 9.1\,(T - T_f)\;. \tag{7.70}$$

The integration in (7.63) furnishes now

$$\ell n\; a_1(T_e) \;=\; -\int_{T_e}^{T_f} \frac{1436 + 9.1(T - T_f)}{RT^2}\,dT$$

$$= \; -\frac{1436}{RT_f}\,q - \frac{9.1}{R}[\ell n(1 + q) - q] \tag{7.71}$$

with the temporary abbreviation

$$q \;=\; (T_f - T_e)/T_e\;. \tag{7.72}$$

Series development of $\ell n(1 + q)$ and conversion to log a_1 lead to

$$\log a_1(T_e) \;=\; -1.1488q + 0.99\,q^2 - 0.66\,q^3 + \dots\;. \tag{7.73}$$

It is customary to introduce the freezing point depression

$$\theta \;=\; T_f - T_e \;=\; qT_e \tag{7.74}$$

and to develop

$$q \;=\; \frac{\theta}{T_f - \theta} \;=\; \frac{\theta}{T_f}\left(1 + \frac{\theta}{T_f} + \frac{\theta^2}{T_f^2} + \dots\right)$$

$$q^2 \;=\; \left(\frac{\theta}{T_f}\right)^2\left(1 + 2\,\frac{\theta}{T_f} + \dots\right)$$

$$q^3 \;=\; \left(\frac{\theta}{T_f}\right)^3 (1 + \dots) \tag{7.75}$$

so that finally we obtain from (7.73)

$$\log a_1(T_e) \;=\; -4.206\cdot10^{-3}\theta(1 + 0.49\cdot10^{-3}\theta \;-\; 2.1\cdot10^{-6}\theta^2). \tag{7.76}$$

In the limit for high dilution Raoult's law tells us that $\gamma_1 = 1$ or $a_1 = x_1 = 1 - x_2$ and

$$\log a_1 = 0.43429\ln(1 - x_2) = -0.43429\, x_2 = -0.004206\,\theta \,. \qquad (7.77)$$

We introduce the molality (Eq.(7.48)) by the approximate relation for high dilution

$$m = 55.506\, x_2 \qquad (7.78)$$

and find

$$\theta = 1.860\, m \,. \qquad (7.79)$$

This is the limiting relation which has been so often used for the determination of the molal weight $\bar{M}_2$ of a solute. Since the weight of solute in 1000 g of solvent is given by

$$w = m\bar{M}_2 \qquad (7.80)$$

we have

$$\bar{M}_2 = 1.860\, w/\theta \,. \qquad (7.81)$$

From $\log a_1(T_e)$ we obtain $\log \gamma_1'$ by subtracting $\log x_1$. The whole calculation is outlined for a few of the freezing point determinations of urea in Table 7.5. Here the first two lines contain a selection of the experimental data of Chadwell and Politi [74.2]. The following three lines show computation indicated by (7.76); the third and fourth lines illustrate the rapid convergence of the power series. The next three lines lead to $\gamma_1' = \log(a_1/x_1)$.The accent is conventional for the activity coefficient at the freezing temperature. The last line shows that $\log \gamma_1'$ is quite well represented by the simplest term, namely, by Bx_2^2 (Eq.(6.47)); this can be only an approximate representation since B varies with the temperature. It

Table 7.5

Freezing Points and Activity Coefficient of Urea Solutions

m	0.3241	0.6458	3.3601	5.2848	8.0828
θ	.5953	1.1698	5.4897	8.0825	11.4142
$0.49\cdot 10^{-3}\theta$	.00029	.00057	.00269	.00396	.00560
$2.1\cdot 10^{-6}\theta^2$	---	---	.00006	.00014	.00027
$-\log a_1(T_e)$	.002505	.004920	.023090	.034125	.048266
x_2	.005805	.011501	.057081	.086934	.127112
$-\log x_1$	.002528	.005024	.025528	.039497	.059039
$\log \gamma_1'$	.000023	.000101	.00238	.00537	.01077
$0.71\, x_2^2$	.000024	.000095	.00231	.00535	.01147

provides, however, a simple computation for the activity coefficient γ_2' of urea, which is now conventionally defined by (7.49). First we calculate the activity coefficient γ_2'' which we define by

$$a_2 = x_2\gamma_2''; \qquad \lim_{x_2=0} \gamma_2'' = 1. \tag{7.82}$$

The choice of a different standard state requires that we have to add a constant c to (6.47) so that we have

$$\log \gamma_2'' = Bx_1^2 + c. \tag{7.83}$$

The constant is eliminated by means of the new standard condition

$$\lim_{x_2=0} \log \gamma_2'' = 0 = B + c; \qquad c = -B$$

so that

$$\log \gamma_2'' = B(x_1^2 - 1) = B(x_2 - 2)x_2 . \tag{7.84}$$

The results are shown in Table 7.6 with B = 0.71.

Table 7.6

Activity Coefficient of Urea

m	0.5	1	3	5	8.08
x_2	.0089	.0177	.0512	.0825	.1267
γ_2''	.971	.944	.850	.773	.678
γ_2'	.963	.928	.806	.709	.592

The switch to the convention for dilute aqueous solutions is based on

$$a_2 = x_2\gamma_2'' = m\gamma_2'; \qquad \lim_{m=0} \gamma_2' = 1, \tag{7.85}$$

which with the aid of (7.48) furnishes γ_2' in Table 7.6.

For the conversion of γ_1' to another temperature T we subtract $\ln x_1$ from Eq.(7.64) and have

$$\log \gamma_1 = \log \gamma_1' - \int_{T_e}^{T} (\bar{L}_1/RT^2)dT. \tag{7.86}$$

We use data given by Gucker and Pickard [74.4] who tabulate extensive power series in m for heat contents, heat capacities and their temperature derivatives. For an approximative illustration we use only the data for 25°C from which we find

$$\bar{L}_1 = 1.548\ m^2 - 0.246\ m^3 + 0.0236\ m^4 - 0.00106\ m^5. \tag{7.87}$$

The results from (7.86) are shown in Table 7.7.

Table 7.7

Activity Coefficient of Water at 25 C in Urea Solutions

x_2	0.05	0.10
m	2.93	6.18
$\bar{L}_1$	8.69	24.0
log γ_1'	0.00177	0.00710
log γ_1 (298)	.00106	.00483
log γ_1 (isopiestic)	.00148	.00458
γ_1 (298)	1.0027	1.0112
γ_1 (isopiestic)	1.0034	1.0106

The choice of molality m for the representation of thermodynamic properties of aqueous solutions has historic reasons. It leads to extended and poorly convergent power series. For non-electrolytes the conversion to mole fractions offers considerable simplifications as shown here for log γ_1'. Similar advantages are expected for calorimetric quantities. The general idea of the activity coefficient is flexible enough that we can switch from one specific definition to another. Conversions of this kind should not present any serious difficulty.

7.44 Solubility. The example of bismuth-cadmium mixtures (Section 7.42) illustrated the fact that there is no essential difference in the information value to be obtained from freezing point temperatures and solubilities as temperature functions. Equation (7.63) applies also to solubility curves. Indicating the solute by subscript 2, and assuming that the heat of solution $\Delta H°$ of the solute is independent of the temperature (equal therefore to its heat of fusion) we may write (7.67) as

$$\log a_2(\text{sat}) = \log x_2(\text{sat}) + \log \gamma_2 = \frac{\Delta H°}{2.303\ R}\left(\frac{1}{T} - \frac{1}{T_f}\right) \qquad (7.88)$$

where a_2(sat) and x_2(sat) mean the saturation values at the temperature T. If $\Delta H°$ varies with the temperature, the difference of the heat capacities is to be taken into account as has been done in (7.71).

Hildebrand has extensively studied solubilities and applied them to the investigation of solutions. Table 7.8 shows a few of the data discussed by Hildebrand and Scott [74.5]. They show the extreme range of imperfection that can be examined by this method.

Table 7.8

Solubility of Iodine in Several Solvents at 25 C

Solvent	x_2(sat)	γ_2
Bromoform	0.0616	4.19
Chloroform	.0215	11.3
Carbon Tetrachloride	.01147	22.5
2,2,3-Trimethylpentane	.00592	43.6
n-Fluoroheptane	.000185	1400

7.5 THE ISOPIESTIC METHOD

The isopiestic method for determining vapor pressures is the simplest one in principle; it offers, moreover, important experimental advantages. It was developed by Sinclair and Robinson [75.1] and greatly refined by Scatchard, Hamer and Wood [75.2]. It is a direct application of the general equilibrium condition: equilibrium between two objects with respect to a generalized coordinate is established if the conjugate generalized forces are equal (Eq.(1.11)). In isopiestic measurements one compares the volatile solvent in samples of solutions of non-volatile solutes by keeping them in a common vapor space and periodically weighing the samples; the coordinate is the number of moles of the solvent, the conjugated force is its partial molal free energy or any monotonically increasing function of it, such as the activity or the partial vapor pressure.

With the proper precautions, thoroughly discussed by Scatchard and his coworkers, the procedure leads to results of high accuracy in a reasonable time. The method is limited, however, by the volatility conditions and the requirement of a set of comparison solutions of known activity.

Two values of log γ_1 interpolated from the results of Scatchard and his coworkers are compared with the freezing point results in Table 7.7. The agreement is satisfactory.

7.6 OSMOTIC PRESSURE

A century ago osmotic pressure was in fashion as a theoretically simple and transparent phenomenon because of its similarity with the law of the perfect gas. The method was more or less forgotten when its accuracy could not be improved to the level of other methods such as freezing point depression. In the last few decades, however, it was revived and successfully applied to the determination of the molal weight and activity of polymers and other high-molecular substances.

In a determination of the osmotic pressure a solution under the elevated pressure $\Pi + 1$ is brought to equilibrium with the pure solvent under atmospheric pressure. This can be achieved by means of a semipermeable membrane which lets pass the solvent but not the solute. The pressure difference Π is called the osmotic pressure. The generalized coordinate is the number of moles n_1; the equilibrium condition prescribes that the decrease of the molal free energy by the presence of the solute must be compensated by the increase of the pressure over the solution. If we indicate elevated pressure in parentheses we have

$$\begin{aligned}\bar{G}_1(\Pi + 1) &= \bar{G}_1 + \Pi(\partial\bar{G}_1/\partial P)_{x_1} + 0.5\,\Pi^2(\partial^2\bar{G}_1/\partial P^2)_{x_1} + \ldots \\ &= \bar{G}_1 + \Pi\bar{V}_1 - 0.5\,\Pi^2\bar{V}_1\beta + \ldots \; . \end{aligned} \tag{7.89}$$

The compressibility

$$\beta = -(1/\bar{V}_1)\cdot(\partial\bar{V}_1/\partial P) \tag{7.90}$$

is of the order of magnitude 10^{-5} atm^{-1}; the last term in (7.89) is therefore negligible.

For the activity we have

$$\ell n\, a_1 = (\bar{G}_1 - \bar{G}_1^\circ)/(RT) = -\Pi\bar{V}_1/RT \; . \tag{7.91}$$

If the solution is perfect we obtain

$$-\ell n\, a_1 = -\ell n(1 - x_2) = x_2 + 0.5\,x_2^2 + \ldots = \Pi\bar{V}_1/RT \; . \tag{7.92}$$

Molal weights of the solutes are calculated by comparison of x_2 and the weight of the solute in 1000 g of the solution in a similar manner as that described in Section 7.43. In order to represent activities of the components one uses preferentially the relations of Scatchard, Hildebrand, Flory, Huggins (Sections 6.34 and 6.35). With substances of high molal weight the volume fractions are much more suitable than the mole fractions.

The principal experimental difficulty of the osmotic method has always been the preparation of a suitable membrane. The development of polymers has added to the early collodion membranes cellophane and various other substances. Above all, the change in objective has brought about an entirely new aspect. Now one does not want membranes that retain all kinds of dissolved molecules: the membrane should retain only molecules of high molal weight and permeability to small molecules is even required. It is the special distinction of the osmotic method that it is not sensitive to impurities of small molecules; they do not contribute to the osmotic pressure because they pass through the appropriately chosen membrane. This advantage is the reason for the revival of the osmotic method. Its usefulness has been increased by considerable experimental improvements [76].

SUMMARY OF THE SEVENTH CHAPTER

Phase equilibria are the main source of quantitative information of solutions. At the same time they are a principal tool in the preparation, purification and analysis of chemical substances.

Equilibria between a liquid solution and its vapor are the foremost source of information, have been investigated in great detail, and are used extensively in distillation. Investigations of the equilibrium between two liquids, and between a liquid solution and a solid have early contributed to our knowledge and technology.

The conditions of stability of a phase with respect to a split into two phases have been discussed in connection with liquid-liquid equilibria. This should not detract from the general validity of these conditions.

The fascinating aspect of a discussion of solutions is the variety of phenomena subject to the same strict laws.

References

(71.1) G. Scatchard, S. E. Wood and J. M. Mochel, J. Phys. Chem., 43 (1939) 119.

(71.2) C. B. Kretschmer and R. Wiebe, J. Amer. Chem. Soc., 71 (1949) 3176.

(71.3) M. A. Rosanoff, C. W. Bacon and J. F. W. Schulze, J. Amer. Chem. Soc., 36 (1914) 1993.

(71.4) W. K. Lewis and E. V. Murphree, J. Amer. Chem. Soc., 46 (1924) 1.
C. G. Boissonnas, Helv. Chim. Acta, 22 (1939) 541.
M. v. Stein and H. Voetter, Z. physik. Chem., 201 (1952) 97.
B. F. Dodge and N. V. Ibl, Chem. Eng. Sci., 2 (1953) 120.
H. C. Van Ness and J. J. Ljunglin, Chem. Eng. Sci., 17 (1962) 531.
N. White and F. Lawson, Chem. Eng. Sci., 25 (1970) 225.

(71.5) O. Redlich, W. E. Gargrave and W. D. Krostek, I & EC Fundamentals, 7 (1968) 211.

(71.6) O. Redlich, A. T. Kister and C. E. Turnquist, Chem. Eng. Progress, Symp. Series 48, No. 2 (1952) 49.

(73.1) O. Redlich and A. T. Kister, J. Chem. Phys., 36 (1962) 2002.
F. J. Ackerman and O. Redlich, J. Chem. Phys., 38 (1963) 2740.
D. W. Hissong and W. B. Kay, AIChE J., 16 (1970) 580.

(73.2) J. P. Kuenen, Z. physik. Chem., 11 (1893) 38; 24 (1897) 667.

(74.1) M. Hansen and K. Anderko, Constitution of Binary Alloys, McGraw-Hill Book Co. New York, 1958. R. R. Hultgren, R. L. Orr, P. D. Anderson and K. K. Kelley, Selected Values of Thermodynamic Properties, J. Wiley, New York, 1963.

(74.2) H. M. Chadwell and F. W. Politi, J. Am. Chem. Soc., 60 (1938) 1291.

(74.3) Lewis-Randall-Pitzer-Brewer, Thermodynamics, 2nd ed. McGraw-Hill Book Co., New York, 1961, p. 407.

(74.4) F. T. Gucker, Jr. and H. B. Pickard, J. Am. Chem. Soc., 62 (1962) 1464.

(74.5) J. H. Hildebrand and R. L. Scott, Solubility of Nonelectrolytes, 3rd ed. Reinhold, New York, 1950.

(75.1) D. A. Sinclair, J. Phys. Chem., 37 (1933) 495; R. A. Robinson and D. A. Sinclair, J. Am. Chem. Soc., 56 (1934) 1830.

(75.2) G. Scatchard, W. J. Hamer and S. E. Wood, J. Am. Chem. Soc., 60 (1938) 3061.

(76) R. U. Bonnar, M. Dimbat and F. H. Stross, Number Average Molecular Weights, Interscience Publishers, New York, 1958.

CHAPTER 8: ELECTROLYTE SOLUTIONS

A considerable part of our treasure of thermodynamic data rests upon observations of electrolytes and their peculiar properties. Moreover, electrolytes have been a fascinating example of interaction between experiment and theory, and of their mutual promotion. A variety of properties are explained, without bias and often quantitatively, by the presence of free electric charges. There may also be some historical interest in the development of the theory of electrolytes; errors and their corrections illustrate the difficulties and pitfalls encountered in the progress of science.

8.1 PROPERTIES OF ELECTROLYTES

The electric conductivity is one of the most characteristic properties of electrolyte solutions. Table 8.1 shows the place of electrolyte solutions in the middle of a wide range extending from metals to dielectrics. The conductivities of concentrated electrolyte solutions and of molten salts are of the same order of magnitude.

Table 8.1

Electric Conductivity
(ohm^{-1} cm^{-1} at room temperature)

Silver		600 000
Iron		100 000
Nichrome		10 000
Silicon		300
Battery acid (H_2SO_4)		0.76
Potassium Chloride	1 M	0.102
	0.01 M	0.0014
Water		10^{-8}
Glass		10^{-10}
Diamond		10^{-14}

Molal weight determinations by the freezing point method (Eq.(7.81)) gave results close to one half of the expected value for salts of the types KCl or $MgSO_4$, one third for $BaCl_2$ and so on. All electrolytes show considerable contraction when dissolved in water or other solvents; some solutions of magnesium salts occupy a smaller volume than the water that they contain.

A very high rate of reaction is typical for electrolytes rather than for solutes in general. The precipitation of silver chloride or barium sulfate proceeds instantaneously; but the hydrolysis of zinc sulfate and other salts is a slow process and the evolution of oxygen from mixtures of permanganate and hydrogen peroxide solutions is rapid only in the presence of manganous salt.

The class character of electrolytes has been early noticed: All cupric salts have a specific blue color, any silver salt mixed in solution with a chloride precipitates silver chloride, and so on. There are exceptions (color change of cupric salts with ammonia, brown color of concentrated cupric bromide solutions) but they are not numerous and are easy to explain. A large number of properties are additive in mixed dilute solutions (freezing point depression, conductivity, volume change, extinction coefficient, refractivity and others).

Ions, electrically charged particles, were generally assumed to be present as intermediates in electrolytic reactions and also in the conductance of the electric current. But the concentration of these ions was assumed to be extremely small, far below any possibility of detection. The revolutionary idea presented by Arrhenius in 1887 was the suggestion that a <u>large part</u> of a dissolved electrolyte is dissociated into ions. The idea sounded very strange, but the successful interpretation of the characteristic electrolyte properties led to its general acceptance in a short time.

In the vicissitudes of electrolyte theory it turned out that the degree of dissociation of many electrolytes was even higher than Arrhenius had found. Later it was sometimes believed that all strong electrolytes are always completely dissociated. Slowly methods were developed that extended the experimental examination of dissociation and other problems of the molecular state to strong electrolytes.

8.2 INDIVIDUAL AND MEAN ACTIVITY COEFFICIENTS

The composition of an electrolyte solution is most frequently indicated by means of the molality m, i.e., as the number of moles of solute per

1000 g solvent (See Eq.(7.48)). In rare cases the mole fraction x_2 is used; this is sometimes more suitable for acids where a continuous set of liquid solutions from high dilution to the pure solute is accessible.

A practically important difference appears in the choice of the standard states of electrolytes and non-electrolytes. We want to define the activity at the standard state as nearly proportionate to the molality in the region of approximate validity of Henry's law. But this law does not hold for m, even in very dilute solutions. It does apply to the ions. So we introduce the molality m_i of an ion i and the "split factor" ν_i by means of

$$m_i = \nu_i m, \tag{8.1}$$

i.e., at high dilution one molecule of the electrolyte produces ν_i ions of species i. (It is unfortunate that the symbol ν_i must be introduced for a quantity that has nothing to do with the reaction coefficients denoted by ν_i in Section 5.1. But the scarcity of available symbols leaves hardly any choice. We must beware of a too rigid link between concept and symbol.)

The partial molal Gibbs free energy $\overline{G}_i$ is now represented by means of the individual activity coefficient γ_i as

$$\overline{G}_i = \overline{G}_i^\circ + RT\ell n(\nu_i m\gamma_i) \tag{8.2}$$

with the convention

$$\lim_{m=0} \gamma_i = 1 , \tag{8.3}$$

which expresses Henry's law.

There is no way to produce a solution which contains only the ion i without "counter ions" to make the solution electrically neutral. But, using the subscript 2 as before for the solute, we have

$$\overline{G}_2 = \Sigma\nu_i\overline{G}_i = \Sigma\nu_i\overline{G}_i^\circ + RT\ \Sigma\nu_i\ell n(\nu_i m\gamma_i) . \tag{8.4}$$

We eliminate not only $\overline{G}_i$ but also $\overline{G}_i^\circ$ and γ_i by means of

$$\overline{G}_2^\circ = \Sigma\nu_i\overline{G}_i^\circ \tag{8.5}$$

$$\nu = \Sigma\nu_i; \qquad \ell n\ \gamma = \Sigma(\nu_i\ell n\ \gamma_i)/\nu . \tag{8.6}$$

For electrolytes we call γ as defined by (8.6) the <u>mean</u> activity coefficient. We obtain

$$\overline{G}_2 = \overline{G}_2^\circ + RT\ell n(m\gamma) + RT\Sigma\nu_i\ell n\ \nu_i. \tag{8.7}$$

If we have only two ionic species we may write

$$\overline{G}_2 = \overline{G}_2^\circ + RT\ell n(\nu_+^{\nu_+}\nu_-^{\nu_-}m^\nu\gamma^\nu). \tag{8.8}$$

Split factors and valences z_i (positive or negative charges of the ions) are connected by the electroneutrality condition

$$\Sigma \nu_i z_i = 0; \qquad \nu_+ z_+ + \nu_- z_- = 0. \tag{8.9}$$

Why do we introduce individual chemical potentials $\bar{G}_i$ and activity coefficients γ_i if we hurry to eliminate them?

First, we obtain real information from Eq.(8.5) although the quantities $\bar{G}_i^o$ cannot be measured and are therefore fictitious. We conclude from (8.4) that

$$\begin{aligned}\bar{G}^o(NaNO_3) + \bar{G}^o(KCl) &= \bar{G}^o(Na^+) + \bar{G}^o(NO_3^-) + \bar{G}^o(K) + \bar{G}^o(Cl) \\ &= \bar{G}^o(NaCl) + \bar{G}^o(KNO_3).\end{aligned} \tag{8.10}$$

In other words, the fictitious quantities $\bar{G}_i^o$ are modules into which we may resolve the measured quantities $\bar{G}^o(NaNO_3)$ and so on, to be used in other combinations.

Secondly, the theory of electrostatic interaction will provide a very useful theoretical representation of individual activity coefficients at low concentrations.

The inspection of a few examples shown in Table 8.2 will help in the tedious job of getting acquainted with these indispensable definitions.

Table 8.2

Split Factors and Valences

Type	$NaNO_3$	$BaBr_2$	$AlCl_3$	$MgSO_4$
ν_+	1	1	1	1
ν_-	1	2	3	1
ν	2	3	4	2
z_+	1	2	3	2
z_-	-1	-1	-1	-2
$\nu_+^{\nu_+}\nu_-^{\nu_-}$	1	4	9	1
a_2	$(m\gamma)^2$	$4(m\gamma)^3$	$9(m\gamma)^4$	$(m\gamma)^2$
μ	m	3m	6m	4m
$(\Sigma\nu_c z_c^2)^{1.5}/(2^{0.5}\nu)$	1	3.464	7.348	8

In accord with (8.8) the activity of the electrolyte is given by

$$a_2 = \nu_+^{\nu_+}\nu_-^{\nu_-}(m\gamma)^\nu . \tag{8.11}$$

Lewis and Randall [82.1] found empirically that the influence of various electrolytes on activity coefficients can be uniformly represented in first approximation as a function of the ionic strength which they define as

$$\mu = 0.5 \sum_i m_i z_i^2 \ . \tag{8.12}$$

The ionic strength presents a common scale for electrolytes of different types. The solubility of thallous chloride, for instance, is higher in solutions of sodium nitrate or potassium sulfate than in water. The increase is the same in solutions of equal ionic strength, not in solutions of equal molality.

In a solution of a single electrolyte the ionic strength is

$$\mu = 0.5m \Sigma \nu_i z_i^2 \tag{8.13}$$

(cf. Table 8.2). But the definition (8.12) covers any mixture.

Shortly after the introduction of the ionic strength its role was interpreted by the theory of Debye and Hückel. In the theory a modified ionic strength based on the molarity c (moles per liter)

$$\mu' = 0.5 \sum_i c_i z_i^2 \tag{8.14}$$

is found to be significant. Since the weight of a solution containing 1 kg solvent and m moles of solute (molal weight $\bar{M}_2$) is $1000(1 + 0.001\ \bar{M}_2 m)$ (gram), the volume concentration or molarity c of this solution is obtained with the aid of the specific weight s as

$$c = ms/(1 + 0.001\ \bar{M}_2 m). \tag{8.15}$$

For dilute solutions we have approximately

$$c = ms^\circ \tag{8.16}$$

with s° being the specific weight of the solvent. In dilute aqueous solutions c and m are approximately equal.

8.3 ELECTROSTATIC INTERACTION

Van Laar was probably the first to point out that the interaction between free electric charges must lead to considerable deviations from the laws of the perfect solution. Similar arguments were proposed early in this century by G. N. Lewis, Wegscheider and Bjerrum.

8.31 The Model. The first quantitative theory was proposed by Malmström [83.1] in 1905. The energy of Coulomb forces in a solution of ions must vary with the concentration because the distances between the ions depend on the concentration. It seems to be plausible that the average distance between ions is proportional to the cubic root of the volume available for one mole of the electrolyte. The reciprocal distance is therefore proportional to the cubic root of the molar concentration, or to $c^{\frac{1}{3}}$. Therefore the electrostatic energy is also assumed to be proportional to $c^{\frac{1}{3}}$.

Malmström's idea did not find any response. (Later it was successfully developed for ionic crystals by Madelung.) But the same model was rapidly and generally adopted when it was again suggested for electrolyte solutions by Ghosh in 1919.

Only Debye realized immediately the defect of this model. A liquid solution of ions differs from a crystal by the mobility of the ions. The average distance between two ions is the result of two counteracting influences, namely, the random thermal motion tending to equalize all distances regardless of the charge, and the Coulomb forces tending to decrease the distance between a positive and a negative ion and to increase the distance between ions of the same sign. Since the thermal energy increases with the temperature, the net influence of the electric force will decrease. This essential feature is not represented by Malmström's model.

The average distance between a positive and a negative ion is always smaller than the average distance between particles without any interacting forces, the average distance between ions of equal signs is larger. The artifice leading from this qualitative idea to quantitative computations is the introduction of a coordinate system that has a certain ion ("central ion") in its origin and travels with this ion. We call its charge $z_c\varepsilon$ with ε representing the electron charge. The essential problem is then: What is the average ionic concentration N_i (number of ions i per ml) of an ion of charge $z_i\varepsilon$ at a distance r from the origin? In a fixed coordinate system the answer would be simply given by the ordinary (stoichiometric or analytical) concentration c_i (moles per liter) and Loschmidt's number per mole N_L, namely,

$$N_i^0 = N_L c_i/1000. \qquad (8.17)$$

But in our moving coordinate system the answer will depend on the distance r; on the average, N_i is higher in the neighborhood of the central ion if z_i and z_c have different signs, and lower if they have the same signs. The ions surrounding a central ion in varying concentrations have been called its ionic atmosphere. The elegant solution of N_i as a function of r has been given by Debye and Hückel [83.2] in 1923 with the aid of the theorems of of Boltzmann and of Poisson.

In applying Boltzmann's theorem (4.26)

$$N_i = N_i^o e^{-u_i} \tag{8.18}$$

we introduce the electrostatic potential ψ at the place of ion i at the distance r from the central ion. The electric energy is $z_i\varepsilon\psi$ and

$$u_i = z_i\varepsilon\psi/kT. \tag{8.19}$$

If all charges disappear, we have $u_i = 0$ and according to (8.18) we obtain $N_i = N_i^o$, i.e., all deviations from uniform distribution disappear; the meaning of N_i^o is therefore the same in (8.17) and (8.18).

Poisson's equation has a purely electrostatic background. It follows from the definition of the potential, namely, that the electric force acting on the unit charge in the direction of x is given by $-\partial\psi/\partial x$. A summation of these forces emanating in all directions from a certain point is given by the behavior of the field ψ in its neighborhood. It is represented by $\partial^2\psi/\partial x^2 + \partial^2\psi/\partial y^2 + \partial^2\psi/\partial z^2$. Often this sum is equal to zero (equation of Laplace); we say then that the point has no free charge. But if the summation does not give zero, we say that

$$\frac{\partial^2\psi}{\partial x^2} + \frac{\partial^2\psi}{\partial y^2} + \frac{\partial^2\psi}{\partial z^2} = -4\pi\rho/D \tag{8.20}$$

thus defining the charge density ρ by this equation of Poisson, where the dielectric constant is denoted by D.

The charge density is given in our model by

$$\rho = \sum_i z_i\varepsilon N_i = \varepsilon\sum_i N_i^o z_i \exp(-z_i\varepsilon\psi/kT). \tag{8.21}$$

The potential field around the central ion is centro-symmetric, i.e., it depends only on r, and not singly on three spatial coordinates. In this case one can transform

$$\frac{\partial^2\psi}{\partial x^2} + \frac{\partial^2\psi}{\partial y^2} + \frac{\partial^2\psi}{\partial z^2} = \frac{1}{r^2}\frac{d}{dr}\left(r^2\frac{d\psi}{dr}\right) = \frac{1}{r}\frac{d^2(r\psi)}{dr^2} \tag{8.22}$$

and obtain

$$\frac{1}{r}\frac{d^2(r\psi)}{dr^2} = -\frac{4\pi\varepsilon}{D}\sum N_i^o z_i \exp(-z_i\varepsilon\psi/kT). \tag{8.23}$$

In the hands of Debye, this differential equation became the key to the electrolyte problem. It had been used before by Laue [83.3] in the early theory of radiotubes.

The Boltzmann-Poisson equation (8.23) determines the electrostatic potential in the environment of the central ion as a function of the distance r from the ion. This is the potential in the environment of a charge $z_c\varepsilon$ in a medium of dielectric constant D, namely,

$$\psi^\circ = z_c\varepsilon/Dr \tag{8.24}$$

modified by the influence of the ionic atmosphere around the central ion. Actually it is this modification, or the part of the potential due to ionic interaction, which is of interest.

In order to find an approximate solution of (8.23), Debye and Hückel develop the exponential function into a power series and retain only the term of first order

$$\frac{1}{r}\frac{d^2(r\psi)}{dr^2} = -\frac{4\pi\varepsilon}{D}\sum N_i^\circ z_i\left(1 - \frac{z_i\varepsilon\psi}{kT} + \ldots\right) . \tag{8.25}$$

It is by no means immediately clear that such a development makes sense. On the contrary, considering the fact that (8.24) is part of ψ and that we shall apply the result to $r = 0$, we should suspect that the series, at the very best, would be extremely poorly convergent. Later extensive studies, however, have justified the bold procedure of Debye and Hückel.

The approximate equation (8.25) becomes very simple if we introduce the electroneutrality condition

$$\sum N_i^\circ z_i = 0 \tag{8.26}$$

for the stoichiometric concentrations. We obtain now

$$\frac{1}{r}\frac{d^2(r\psi)}{dr^2} = \frac{4\pi\varepsilon^2}{DkT}\sum N_i^\circ z_i^2\cdot\psi . \tag{8.27}$$

We multiply this equation by r, introduce (8.14) and (8.17), and abbreviate

$$\kappa^2 = \frac{4\pi\varepsilon^2}{DkT}\sum N_i^\circ z_i^2 = \frac{8\pi\varepsilon^2}{DkT}\cdot\frac{N_L}{1000}\cdot\mu' . \tag{8.28}$$

Thus we have

$$\frac{d^2(r\psi)}{dr^2} = \kappa^2(r\psi) . \tag{8.29}$$

The great power of this approximate equation rests on its simplicity. It contains three kinds of parameters:

(a) universal constants, namely, ε, k, N_L;

(b) the dielectric constant D of the solvent;

(c) the concentrations c_i and charges z_i of the ions.

All these quantities, and also the temperature, are known or experimentally given. The interaction effect is described without any arbitrary parameter.

Equation (8.29) is well known in the theory of harmonic motion. Its solution

$$\psi = [A\exp(-\kappa r) + A'\exp(\kappa r)]/r \tag{8.30}$$

is easily verified. Of the two integration constants A' is zero because the potential disappears for large values of r. For infinite dilution ($\mu' = 0$; $\kappa = 0$) interaction disappears so that the potential has the special value (8.24)

$$\psi^{o} = A/r = z_c\varepsilon/(Dr). \tag{8.31}$$

The integration constant is therefore

$$A = z_c\varepsilon/D \tag{8.32}$$

and the potential is

$$\psi = (z_c\varepsilon/Dr)\exp(-\kappa r) = z_c\varepsilon/D[1/r - \kappa + 0.5\,\kappa^2 r - \ldots] . \tag{8.33}$$

The potential of interaction is therefore in first approximation

$$\psi - z_c\varepsilon/(Dr) = -(z_c\varepsilon\kappa/D)(1 - 0.5\,\kappa r + \ldots) . \tag{8.34}$$

It is a remarkable result of (8.34) that the interaction potential has at the place of the central ion ($r = 0$) a finite value, namely,

$$\Phi = -z_c\varepsilon\kappa/D \tag{8.35}$$

which, moreover, can be calculated from the known properties of the ions and the solvent.

The interaction potential (8.35) is always small compared with the self-potential $-z_c\varepsilon/(Dr)$ of the central ion. This is illustrated in Table 8.3,

Table 8.3

Average Concentrations in the Ionic Atmosphere of K^+ in 0.01 M KCl (grams-ions per liter)

Distance (Å)	0	1	3	10	30	100
c'_-	-	(0.0203)	0.0127	0.0107	0.01024	0.01007
c'_+	-	(.0049)	.0079	.0093	.00976	.00993
c''_-	-	(.0198)	.0124	.0105	.01009	.01002
c''_+	-	(.0051)	.0081	.0095	.00991	.00998
$c'_-/c''_- = c''_+/c'_+$	(1.026)	(1.026)	1.022	1.020	1.015	1.007

which shows its effect on the ion distribution around the central ion. The computation is based on Eqs.(8.18), (8.19), (8.24) and (8.33). The distance r from the origin is indicated in Å (Ångström units, 10^{-8} cm). The concentration of Cl^- is indicated by c'_- if only the self-potential is taken into account and the interaction potential is disregarded; it is always greater than the stoichiometric concentration; the concentration c'_+ of the K^+ ions is always smaller.

Both concentrations are only slightly modified from c'_- and c'_+ to c''_- and c''_+ if the interaction potential is taken into account, i.e., if the concentrations are calculated with the full potential (8.33). Yet it is the inconspicuous change in concentrations from c'_-, c'_+ to c''_-, c''_+ which determines the appreciable effect found in the thermodynamic properties.

Since the radii of the ions amount to more than 1 Å, the figures put in parentheses in Table 8.3 have no immediate significance.

8.32 Activity Coefficients. The primary object of this discussion is the contribution made by ionic interaction to the Gibbs free energy of an ion. The interaction between an ion of species c and the other ions is given by the forces between the central ion and its ionic atmosphere. The contribution to the free energy is obtained (Eq.(2.39)) by calculating the (reversible, isothermal, isobaric) work W required to put the charge $z_c\varepsilon$ on the central ion against the action of the surrounding ions; we do not take into account the work against the self-potential of the ion since this term is implied also in the standard data of the ion and therefore cancels automatically when we proceed from the partial molal free energy to the activity coefficient. The interaction contribution of N_L ions of species c is then contained in the Gibbs free energy (8.2) as

$$N_L W = RT\ell n\, \gamma_c\ . \tag{8.36}$$

There are various ways for deriving W from Eq.(8.35). The most direct one has been suggested by Güntelberg. We freeze the positions and charges of all ions in the ionic atmosphere; the quantity κ in (8.35) therefore stays constant. Then we start with a central ion from which we have removed its charge, and charge it (reversibly, isothermally, isobarically) to its proper charge $z_s\varepsilon$. The instantaneous charge during this process we call $\lambda z_s\varepsilon$ with λ going from 0 to 1. The work done upon the central ion is

$$W = -\int_{\lambda=0}^{\lambda=1} (\lambda z_c \varepsilon\kappa/D)\, d(\lambda z_c\varepsilon) = -0.5 z_c^2\varepsilon^2\kappa/D. \tag{8.37}$$

If we repeat this process for every ion in the solution, we have charged all ions precisely once to their proper potential.

For the activity coefficient we obtain from the last two equations

$$\log \gamma_c = N_L W/(2.303\ RT) = W/(2.303\ kT)$$
$$= -\frac{0.5\ \varepsilon^2\kappa}{2.303\ DkT}\cdot z_c^2\ . \tag{8.38}$$

The mean activity coefficient (Eq.(8.6)) is therefore

$$\log \gamma = -\frac{0.5\ \varepsilon^2 \kappa}{2.303\ DkT} \cdot \frac{\Sigma \nu_c z_c^2}{\nu} \quad . \tag{8.39}$$

We abbreviate

$$h = \left(\frac{2\pi N_L}{1000}\right)^{0.5} \cdot \frac{\varepsilon^3}{2.303(DkT)^{1.5}} \quad , \tag{8.40}$$

introduce the valence factor

$$w = 0.5\ \Sigma \nu_c z_c^2 \ , \tag{8.41}$$

and obtain the mean activity coefficient

$$\log \gamma = -2wh\sqrt{\mu'}/\nu \ . \tag{8.42}$$

If only two kinds of ions are present we may replace μ' by means of (8.14) and (8.16) so that

$$\log \gamma = -2w^{1.5}hc^{0.5}/\nu = -2w^{1.5}h(s^\circ m)^{0.5}/\nu \ . \tag{8.43}$$

For non-aqueous solutions the factor s° deviates usually appreciably from unity so that it must not be disregarded; for aqueous solutions it is practically equal to unity.

The great power of this relation as well as its limitations are illustrated in Figs. 8.1 and 8.2. The diagrams show log γ at 25°C as a function of $m^{0.5}$. The calculated values (8.43) are indicated by broken straight lines. They furnish excellent limiting slopes. The great influence of the valences z_c of the ions is in perfect agreement with the observed data (Fig. 8.1). The difference between the two curves for HCl in water (D = 78.8) and in aqueous methyl ethyl ketone (20 wt %, D = 74.9 for the mixture) is due to the influence of the dielectric constant in (8.40). It is accurately represented by the limiting slopes.

But the deviations from the limiting slopes are noticeable for such low concentrations as m = 0.01. For the higher valence types the deviations are even quite large. It is strictly only the limiting slopes that can be calculated for any temperature from the dielectric constant of the solvent and the valence type of the electrolyte.

From the beginning Debye and Hückel tried to represent the deviations from the limiting law by taking into account the individual sizes of the ions. The resulting relations indeed extended the range of agreement and could represent such individual differences as between NaCl and KNO_3 in Fig. 8.1. In view of hydration of the ions one could not expect agreement of these ionic radii in solutions with the values found in crystals. But the usefulness of the relations containing the ionic radius has been questionable because the

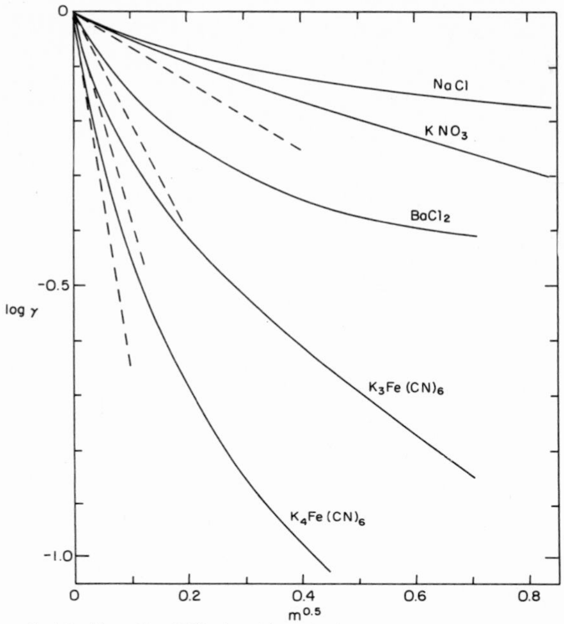

Fig. 8.1. Activity Coefficients of Strong Electrolytes (The broken lines represent the limiting law of Debye and Hückel. Data: Landolt-Börnstein, Tables 5th ed., 3rd Suppl., p. 2138, Springer, Berlin,1935)

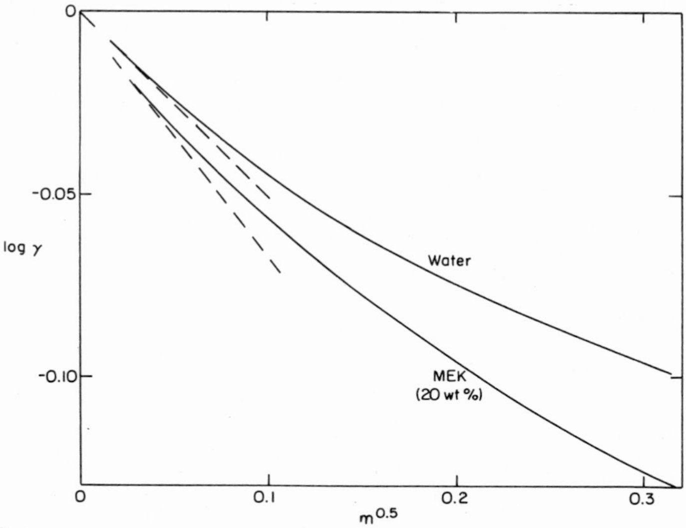

Fig. 8.2. Hydrogen Chloride in Water and in Aqueous Methyl ethyl ketone (25 C. Data: D. Feakins and C. M. French, J. Chem. Soc. 1957, 2284)

radius has remained an empirical parameter to be derived from the observed data. The very great effort of numerous authors to expand the validity of the theory led only to partial success.

The theory, nevertheless, had a great and very beneficial influence. It clarified the deviations from the perfect solution and brought reasonable order into an utterly confused situation. Practically the predicted limiting slopes enables us to extrapolate safely to infinite dilution and to establish the conventional standard state with little loss of the accuracy of the observed data.

8.33 Heat Content and Volume. Whenever a Gibbs free energy or a contribution to the free energy is given as a function of temperature and pressure, we know also the corresponding terms of all other thermodynamic functions (Section 2.4). The electrostatic interaction term of the heat content of an electrolyte was derived from the theory of Debye and Hückel by Gross and Halpern [83.4] and a little later by Bjerrum [83.5].

The interaction term of the partial molal heat content $\bar{H}_2^I$ of a dissolved electrolyte is obtained by summing the corresponding terms for the ions

$$\bar{H}_2^I = \sum_c \nu_c \bar{H}_c \qquad (8.44)$$

or according to (6.42), (8.6) and (8.43) as

$$\bar{H}_2^I = -RT^2 \sum_c \nu_c (\partial \ell n\, \gamma_c/\partial T)_m = -2.303\, RT^2 \nu (\partial \log \gamma/\partial T)_m$$

$$= 2.303\; 2RT^2 w^{1.5} (s^\circ m)^{0.5} (\partial h/\partial T)_m \qquad (8.45)$$

if we disregard a small contribution due to the temperature dependence of the specific gravity s° of the solvent. From (8.40) we obtain

$$\frac{\partial h}{\partial T} = -\frac{1.5h}{DT} \cdot \frac{\partial (DT)}{\partial T} \quad . \qquad (8.46)$$

Since the interaction contribution disappears for high dilution it is equal to the relative partial molal heat content

$$\bar{L}_2 = \bar{H}_2^I - \bar{H}_2^\circ = -2.303 \cdot 3(RT/D) w^{1.5} c^{0.5} \cdot \partial (DT)/\partial T. \qquad (8.47)$$

This relation has been well confirmed by accurate measurements as a limiting law. Another differentiation of (8.44) furnishes the interaction term for the heat capacity.

It was realized later [83.6] that the theory of Debye and Hückel can be used to derive an interaction term for the volume. Indeed, from (3.10), (6.31) and (8.43) we derive for the interaction part $\bar{G}_2^I$ of the partial molal Gibbs free energy the relation (similar to 8.45)

$$\left(\frac{\partial \bar{G}_2^I}{\partial P}\right)_{T,m} = \bar{V}_2^I = RT\,\nu\,\frac{\partial \ln \gamma}{\partial P} = -2.303\cdot 2RTw^{1.5}m^{0.5}\left(\frac{\partial(h\sqrt{s^\circ})}{\partial P}\right)_{T,m} \quad (8.48)$$

implying, as before, that the deviation of γ from unity is determined entirely by the ionic interaction. The relative pressure dependence of s° is of the same order of magnitude as that of h and therefore by no means negligible. We introduce the compressibility β of the solvent

$$\beta = -\frac{1}{\bar{V}^\circ}\left(\frac{\partial \bar{V}^\circ}{\partial P}\right)_T = \frac{1}{s^\circ}\cdot\frac{\partial s^\circ}{\partial P} \quad (8.49)$$

and derive from (8.40)

$$\frac{\partial h}{\partial P} = -\frac{1.5h}{D}\cdot\frac{\partial D}{\partial P} = -1.5\,\frac{\partial \ln D}{\partial P} \quad . \quad (8.50)$$

Thus we obtain

$$\bar{V}_2^I = 2.303\cdot 3RT\, w^{1.5}h(ms^\circ)^{0.5}(\partial \ln D/\partial P - \beta/3). \quad (8.51)$$

Assuming that the deviation of $\bar{V}_2$ from the constant limiting value $\bar{V}_2^\circ$ for high dilution is fully expressed by the interaction term $\bar{V}_2^I$ we have

$$\bar{V}_2 = \bar{V}_2^\circ + 1.5\,k\,w^{1.5}\,(ms^\circ)^{0.5} \quad (8.52)$$

with the abbreviation (not to be confused with Boltzmann's constant)

$$k = 2.303\cdot 2RT\,h\,(\partial \ln D/\partial P - \beta/3)$$
$$= N_L^{\,2}\varepsilon^3[8\pi/(1000\,D^3RT)]^{0.5}(\partial \ln D/\partial P - \beta/3). \quad (8.53)$$

Since the work of Marignac and Traube it has been customary to express variations in the molal volume of solutions by means of the apparent molal volume ϕ defined by the representation

$$V = n_1V_1^\circ + n_2\phi = 1000/s^\circ + m\phi \quad (8.54)$$

of the total volume V. The idea is to split the volume V into a constant contribution $1000/s^\circ$ of the solvent and a variable term $m\phi$, which is thus defined by

$$\phi = (V - 1000/s^\circ)/m \ . \quad (8.55)$$

What actually takes place in a solution is the opposite: The volume of the solute, as far as we know, is approximately constant (except for hydration or solvation) and the changes occur mainly in the solvent. The conventional defintion (8.55), though not based on a valid molecular interpretation, is nevertheless admissible and in general use.

From (8.54) we derive the partial molal volume of the solute

$$\bar{V}_2 = (\partial V/\partial m)_{T,P} = \partial(m\phi)/\partial m \quad (8.56)$$

so that

$$m\phi = \int_0^m \bar{V}_2 dm. \tag{8.57}$$

The theory predicts therefore, according to (8.52),

$$\phi = \bar{V}_2^o + k\,w^{1.5}\,(ms^o)^{0.5} = \bar{V}_2^o + K\,w^{1.5}c^{0.5}\ . \tag{8.58}$$

The similarity of the relations (8.43), (8.47) and (8.58) is conspicuous.

For various strange reasons, Eq.(8.58) was questioned for a long time, although even early data [83.7], shown in Figs. 8.3 and 8.4, did not leave

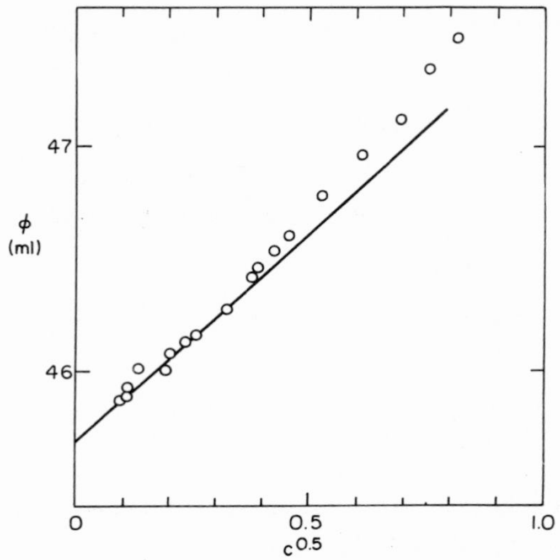

Fig. 8.3. Molal Volume of Potassium Chlorate (25 C. The straight line represents the limiting slope)

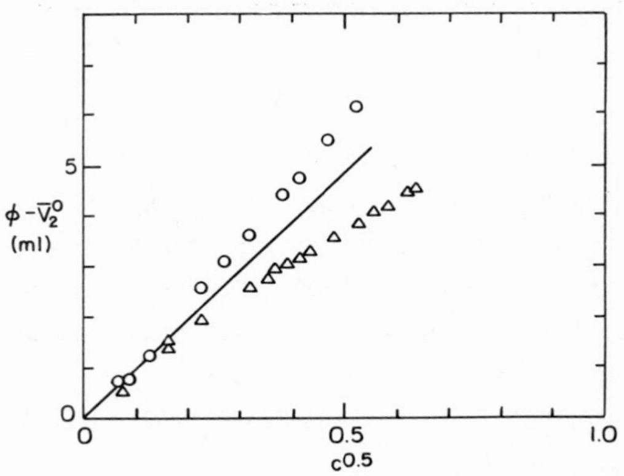

Fig. 8.4. Increase of the Molal Volume (25 C. o sodium sulfate; Δ strontium chloride)

room for any reasonable doubt. Some authors did not realize that an accuracy of at least 10^{-6} in density measurements was required for a valid test. The pressure coefficient $\partial D/\partial P$ of the dielectric constant of water, which is difficult to measure, caused some uncertainty. Later measurements of the dielectric constant removed any possible doubt. The calculated values of the coefficient k have recently [83.8] been confirmed in the range 0-65 C.

The interpretation of the concentration dependence of the molal volumes has been useful in investigations of the dissociation of moderately strong electrolytes and of the hydration of ions.

Another differentiation of (8.58) leads to apparent compressibilities.

8.34 Mixtures. In any reasonable expansion of electrolyte properties the coefficient of $c^{0.5}$ is established; but the coefficients of c and higher powers are individual and empirical. Useful general rules for the thermodynamic properties of electrolyte mixtures have been proposed by Young [83.9] and his coworkers, and repeatedly discussed in recent years.

8.4 GALVANIC CELLS

From the pacemaker ensuring the function of a heart to equipment for emergency lighting there are numerous situations in modern life that require a reliable and convenient energy source. Galvanic cells satisfy the need wherever the power required is small, as in a wrist watch, or the cost is unimportant, as in space equipment, or the weight is a minor concern, as in an automobile starter.

Beyond their immediate applications, galvanic cells are an important source of information in chemical thermodynamics. If we did not have this chemical interest we might represent a cell as an electrical instrument quite similar to a capacitor (as a "black box"). Its state would be indicated by the electric charge as a generalized coordinate, and energy, entropy and other thermodynamic functions could be introduced in the usual manner.

Actually we describe a cell usually by stating the numbers of moles of each component present in each phase, and by stating the reaction tied to the charge transfer from one electrode to the other. The changes in the numbers of moles are linked by the stoichiometric conditions, and the charge transfer is given by Faraday's Law. An example will be more useful than an abstract discussion.

8.41 <u>The Lead Storage Cell</u>. The lead storage cell is an example of a close link between thermodynamic theory and a variety of practical applications. Some types of lead cells are designed for very low discharge rates; cells of this kind come very close to an "ideal cell," i.e., one in which a single, well-defined reaction takes place in either direction. Such cells are very different from the type that can deliver currents of 100 amp. for a short time as required in automobiles.

One electrode of the lead storage cell is essentially lead, the other lead dioxide; the electrolyte is a solution of sulfuric acid. Using the notation of Sections 5.1 and 5.3, we describe the reaction by

$$Pb(s) + PbO_2(s) + 2H_2SO_4(sol'n) = 2PbSO_4(s) + 2H_2O(sol'n) \, . \qquad (8.59)$$

j	1	2	3	4	5
ν_j	-1	-1	-2	2	2

If we mix the reactants, this reaction proceeds spontaneously. However, if we keep lead dioxide and lead separated from each other, the cathode reaction

$$Pb(s) + H_2SO_4(sol'n) = PbSO_4(s) + 2H^+ + 2e^- \qquad (8.60)$$

and the anode reaction

$$PbO_2(s) + 2H^+ + H_2SO_4(sol'n) + 2e^- = PbSO_4(s) + 2H_2O(sol'n) \qquad (8.61)$$

cannot proceed because free electrons e^- cannot accumulate at the cathode and are not available at the anode as long as the electrodes are disconnected. But if we connect them by an electrical conductor, the electrons flow through it from the cathode to the anode. The cell is short-circuited and the reaction proceeds spontaneously.

In charging, the process can be forced in the opposite direction by means of an external force driving electrons from the anode outside to the cathode. The generalized force, conjugate to the charge, is the electromotive force or the voltage. The minimum voltage required to force the reaction (8.59) in the charging direction (from right to left) is the reversible voltage E. If the electrodes are connected by an external voltage smaller than E, the reaction proceeds in the spontaneous direction and the cell is discharged.

Equations (8.60) and (8.61) show that two electrons are transferred if the amounts specified in (8.59) react. The charge transferred is therefore $2\mathsf{F}$ if we express the Faraday equivalent by Loschmidt's number per mole N_L and the electron charge ε as

$$\mathsf{F} = N_L \varepsilon \, . \qquad (8.62)$$

The work done upon the cell by reversible charging is given by Eq.(2.37) as 2FE; for the discharge the reversible work done upon the cell is -2FE and this is the change in the Gibbs free energy for an isothermal-isobaric process (8.59)

$$\Delta G = -2FE \qquad (8.63)$$

where the symbol Δ is used as in Section 5.3. In an irreversible discharge, from the <u>same</u> initial to the <u>same</u> final state, the change ΔG is the same but the work delivered to the outside (e.g., to the starter) and the voltage are smaller. On charging, $-\Delta G$ again is unchanged, but it is smaller than the work we have to put in (we have to increase the voltage above E).

The sign of E, positive in this example, is by convention chosen opposite to that of ΔG.

The lead cell presents a good reminder that the relation $\Delta G = 0$ is <u>not</u> the <u>general</u> equilibrium condition but only the equilibrium condition if <u>no useful work</u> can be done in an isothermal isobaric process.

If we resolve (8.63) as prescribed in (5.14) we obtain

$$\Delta G = 2G_4 + 2\bar{G}_5 - \bar{G}_1 - \bar{G}_2 - 2\bar{G}_3 = -2FE. \qquad (8.64)$$

The solid substances are in the standard states; for water we introduce its activity a_5, as given by (6.29) and (6.30); for sulfuric acid we introduce the molality m_3 and the mean activity coefficient γ_3 (Section 8.2). Thus we obtain

$$\Delta G = 2\bar{G}_4^0 + 2[\bar{G}_5^0 + RT\,\ell n\, a_5] - \bar{G}_1^0 - \bar{G}_2^0 - 2[\bar{G}_3^0 + RT\,\ell n(4m_3^3\gamma_3^3)]$$

$$= \Delta G^o + 2RT\,\ell n\, a_5 - 2RT\,\ell n(4m_3^3\gamma_3^3)\ . \qquad (8.65)$$

8.42 <u>Half-Cells</u>. The advantage of considering half-reactions such as (8.60) and (8.61) is the possibility of various combinations. We can combine not only these two half-reactions with each other but also either one with other half-reactions such as

$$0.5\ H_2(g) = H^+ + e^- \qquad (8.66)$$

$$2Hg + H_2SO_4(sol'n) = Hg_2SO_4(s) + 2H^+ + 2e^-. \qquad (8.67)$$

By measuring the thermodynamic quantities involved in a moderate number of half-reactions we gain information covering a very large number of reactions. The actual assembly of two electrodes characterized by such half-reactions is important because of the data that can be obtained by measuring the reversible electromotive force of the complete cells.

The general condition for the construction of a cell for measurement is the actual realization of the intended reaction under conditions of reversibility and with exclusion of any side reaction. This implies that the changes on both electrodes must proceed reversibly and exclusively, and that the linkage between the two electrodes must not cause any additional change.

These conditions can be satisfied in any of the combinations of the electrodes discussed in these examples. However, there are restrictions. The mercury-mercurous sulfate electrode cannot be used in highly dilute sulfuric acid because there mercurous sulfate hydrolyzes so that the product is a basic sulfate, which is not in accord with the prescribed reaction (8.67). Irreversible changes, such as mixing, are avoided by having the same electrolyte (sulfuric acid of the same concentration) in both half-cells. But this condition is not absolutely strict: In a combination of (8.61) and (8.66) hydrogen must not be in contact with lead dioxide because it would cause an irreversible reaction; the difference, however, between hydrogen-free and hydrogen-saturated sulfuric acid is so small that no disturbing potential difference exists at the interface.

We cannot measure, in a strict sense, the free energy change of any half-reaction. For convenience in data recording, the change in free energy (heat content, heat capacity) of the reaction (8.66) at the hydrogen electrode is set at zero. This reaction proceeds reversibly at an electrode covered with finely divided platinum or palladium in most acid solutions and in dilute alkaline solutions.

Latimer suggested to write half-reactions always in such a manner that n electrons appear at the right of the equality sign (direction of oxidation; the reaction (8.61) would be reversed). One can tabulate either ΔG° or

$$\mathbf{E} = -\Delta G^\circ/(nF). \qquad (8.68)$$

These values are called standard oxidation potentials. They are values computed for all components being in the standard states (unit activity). An abundance of data can be found in Latimer's book [84.1].

One may, of course, write down electrode potentials for substances that are not in the standard state. Thus we write for (8.66), denoting the partial pressure of hydrogen by p_{H_2},

$$\Delta G = \Delta G^\circ + RT\ell n(m_{H^+}\gamma_{H^+}) - 0.5\ RT\ell n\ p_{H_2} \qquad (8.69)$$

assuming that the gaseous phase is perfect. Correspondingly we have to write

$$0.5\ H_2(g) = H^+ + e^-; \qquad \mathbf{E} = E^\circ - (RT/F)\ell n(m_{H^+}\gamma_{H^+}/p_{H_2}{}^{0.5}). \qquad (8.70)$$

For (8.67) we write

$$E = E^\circ - (RT/2F)\ell n(m_{H^+}^2\gamma_{H^+}^2/a_{H_2SO_4}) \tag{8.71}$$

and so on.

Numerous half-cells have been constructed and their potentials have been derived from measurements of suitable combinations. Metals often give well-defined potentials in solutions of their ions; many reactions involving non-metals are also useful. A few examples are shown in Table 8.4

Table 8.4

Standard Potentials of Half-Reactions at 25 C

			E°
$Li(s)$	=	$Li^+ + e^-$	2.9578
$Na(s)$	=	$Na^+ + e^-$	2.7125
$Zn(s)$	=	$Zn^{++} + 2e^-$	0.763
$Fe(s)$	=	$Fe^{++} + 2e^-$	0.440
$Pb(s)$	=	$Pb^{++} + 2e^-$	0.122
$H_2(g)$	=	$2H^+ + 2e^-$	0.0000
$Cu(s)$	=	$Cu^{++} + 2e^-$	-0.337
$Cu(s)$	=	$Cu^+ + e^-$	-0.521
$Ag(s)$	=	$Ag^+ + e^-$	-0.7995
$2I^-$	=	$I_2(s) + 2e^-$	-0.5357
$2Cl^-$	=	$Cl_2(g) + 2e^-$	-1.3594
$Hg + Cl^-$	=	$0.5\ Hg_2Cl_2(s) + e^-$	-0.2700
$Ag + Cl^-$	=	$AgCl(s) + e^-$	-0.2245

The potential series of the metals has a very direct significance; any more positive metal (e.g., sodium) can precipitate any less positive metal (copper) from a solution of its ions.

It is important, however, to realize that it is the change of free energy that combines by addition, not the potential. In order to find the standard potential of ferric ion we may start from the standard potential of ferrous ion

$$Fe = Fe^{++} + 2e^-; \qquad E_a^\circ = 0.440 \tag{8.72}$$

and from information on the equilibrium between ferrous and ferric salts

$$Fe^{++} = Fe^{+++} + e^-; \qquad E_b^\circ = -0.771. \tag{8.73}$$

We cannot find the potential E_c° of the ferric ion by addition of E_a° and E_b° because the corresponding arrangement of the two half-cells would require that electrons are taken out from both electrodes. But going back to free energy changes we have

$$\Delta G_a^\circ = -2FE_a^\circ = -0.880\ F \tag{8.74}$$

$$\Delta G_b^\circ = -FE_b^\circ = 0.771\ F \tag{8.75}$$

$$Fe = Fe^{+++} + 3e^-; \qquad G_c^\circ = -3FE_c = \Delta G_a^\circ + \Delta G_b^\circ \tag{8.76}$$

so that (Luther's rule)

$$E_c = -(\Delta G_a^\circ + \Delta G_b^\circ)/3 = (2E_a^\circ + E_b^\circ)/3 = 0.036. \tag{8.77}$$

We have used in(8.73) the potential of an equilibrium reaction between two ions. This and similar potentials have been actually measured. For the ferrous-ferric equilibrium one determines the electromotive force of a cell containing a platinum electrode in a solution of ferrous and ferric chlorides and as second electrode a silver wire covered with silver chloride in a chloride solution (potential in Table 8.4).

8.5 DETERMINATION OF ACTIVITY COEFFICIENTS

A variety of information can be derived from thermodynamic data of electrolyte solutions. The methods of measuring activity coefficients, therefore, have been highly developed. Several methods, each one limited in its application, complement one another.

The activity a_1 of the solvent and the mean activity coefficient γ of a dissolved electrolyte are linked by the relation of Gibbs, Duhem and Margules (Eqs.(1.31),(6.29),(6.30),(8.8)). The definition of the molality

$$m = 1000\ x_2/(\bar{M}_1 x_1) \tag{8.78}$$

($\bar{M}_1$ = molal weight of the solvent) furnishes

$$x_1 = 1/(1 + 0.001\ \bar{M}_1 m); \qquad x_2 = 0.001\ \bar{M}_1 m/(1 + 0.001\ \bar{M}_1 m) \tag{8.79}$$

so that the relation can be written

$$d \log a_1 + 0.001\ \bar{M}_1 \nu m\ d \log(m\gamma) = 0. \tag{8.80}$$

In the range where the approximation of Debye and Hückel is valid, we derive from (8.43)

$$\begin{aligned} d \log a_1 &= -0.001\ \bar{M}_1 \nu m\ d \log(m\gamma) \\ &= 0.001\ \bar{M}_1[-0.43429\nu dm + hw^{1.5}(s^\circ m)^{0.5} dm]. \end{aligned} \tag{8.81}$$

Integration furnishes

$$\log a_1 = 0.001\ \bar{M}_1[-0.43429\nu m + 0.6667\ hw^{1.5}(s^\circ m^3)^{0.5}]. \qquad (8.82)$$

The first term in the brackets expresses Raoult's law, the second the deviation produced by ionic interaction; it is linear in $m^{1.5}$ while the first deviation in non-electrolyte solutions would be approximately represented by a term containing m^2.

8.51 Freezing Points. Since freezing point depressions have been determined within 0.0001°C or even better, the method has been the most accurate one for activity coefficients in dilute aqueous solutions. The experimental problem has been the establishement of a well-defined phase equilibrium rather than the sensitivity of the thermometer.

The computation of the activity a_1 of the solvent, as outlined in Section 7.41 and the first part of Section 7.43 up to Eq.(7.76) is not changed by dissociation of the solute or ionic interaction since the derivation proceeds in purely thermodynamic terms. But Raoult's law is based on the actual number of molecules and ions present. By comparing (7.76) with the approximation (8.82) we obtain the limiting law

$$\theta/m = 1.860\ \nu(1 - 1.535\ hw^{1.5}c^{0.5}/\nu)/(1 + 0.049\ \theta - 2.1\ 10^{-6}\theta^2). \qquad (8.83)$$

If this relation is compared with (7.79), the factor ν is seen to represent the influence of dissociation and the term containing h the electrostatic effect.

For non-electrolytes the deviation of θ/m from the limiting value would be represented by a power series in x_2 (or m or c). The Debye-Hückel term with the exponent 0.5 extends its influence to lower concentrations than the first term of a power series in c. This is a characteristic feature of ionic interaction.

For the computation of activity coefficients one obtains $\log a_1$ for the freezing temperatures and converts to the desired temperature by the procedure of Sections 7.41 and 7.43. The activity coefficient γ of the electrolyte is then calculated by an integration of (8.80). Lewis and Randall and several later authors [85.1] devised sophisticated methods for this operation; the discontinuity of (8.80) for m = 0 was circumvented by introducing the function

$$j = 1 - \theta/(1.860\ \nu m) \qquad (8.84)$$

instead of θ.

8.52 <u>Electromotive Forces</u>. If we combine a hydrogen electrode and a silver-silver chloride electrode (Table 8.4) we obtain a cell whose reaction is

$$0.5\ H_2(g) + AgCl(s) = Ag(s) + HCl(m); \qquad E^\circ = 0.2245. \tag{8.85}$$

The electromotive force of this cell is given as a function of the molality m of hydrogen chloride and the partial pressure p of hydrogen by

$$E = -\Delta G/F = E^\circ - (RT/F)\ln(m^2\gamma^2/p^{0.5}). \tag{8.86}$$

The factor to be used with a decadic logarithm at 25 C is

$$2.303\cdot RT/F = 0.059158\ v \tag{8.87}$$

so that

$$E = E^\circ - 2\cdot 0.059158\ \log(m\gamma/p^{0.25}). \tag{8.88}$$

The voltage of a reversible cell changes by 0.059158 v if the activity of an ion or the fugacity of a gas (participating in the cell reaction in the first order) changes by a factor of 10.

A series of measurements of cells of the type (8.85) can furnish very accurate values of activity coefficients. The value of E°, in general, cannot be assumed to be known beforehand. The computation problem is therefore the extrapolation of the observed values of E to m = 0; then we make use of the convention

$$\lim_{m=0} \gamma = 1.$$

If measurements have been made at different partial pressures of hydrogen, one corrects for the deviations. The function

$$\log \gamma' = -0.5\ E/0.059158 - \log m = -0.5\ E^\circ/0.059158 + \log \gamma \tag{8.89}$$

can then be computed for every measured value E. It is plotted against $m^{0.5}$. The limiting slope is predicted by the relation (8.43) of Debye and Hückel. This procedure is shown for hydrogen chloride in Fig. 8.5 in a selection of experimental data at 25 C.

The extrapolation of log γ' to m = 0 furnishes -1.879. Since log γ = 0 for m = 0, the extrapolated value of log γ' furnishes the result

$$E^\circ = 2\cdot 0.059158\cdot 1.879 = 0.2245 \tag{8.90}$$

as shown in Table 8.4. The activity coefficients in the experimental range are now given as

$$\log \gamma = \log \gamma' + 1.879. \tag{8.91}$$

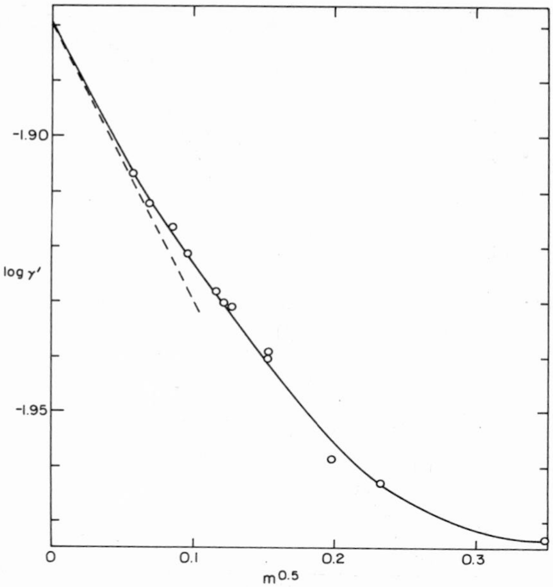

Fig. 8.5. Hydrogen Chloride (25 C. EMF data: R. W. Ehlers and H. S. Harned, J. Am. Chem. Soc., 54(1932)1350)

Numerous activity coefficients of electrolytes in aqueous and non-aqueous solutions have been determined by means of galvanic cells. The hydrogen chloride cell is outstanding in regard to reversibility and absence of side reactions. But the electrodes of mercury, zinc, cadmium, copper, bromide, iodide, sulfate have also been used for accurate measurements.

8.53 Vapor Pressure. The isopiestic method (Section 7.5) has for electrolyte solutions the same advantages and limitations as for non-electrolytes. The computation is changed inasmuch as the relation of Gibbs and Duhem is now expressed by (8.80) and the limiting laws of Debye and Hückel are advantageously used in the same way as in freezing point calculations. The method fails if the solute has a measurable vapor pressure of its own.

When no other method can be applied there remains the measurement of the vapor pressure. The activity coefficients of nitric acid have been derived from freezing points [85.2] up to m = 6.6. No suitable galvanic cell has been found. Vapor pressures have been repeatedly measured. The best determinations [85.3] barely overlap with the freezing points. Figure 8.6 combines freezing point and vapor pressure results [85.4]. The activity coefficient is here

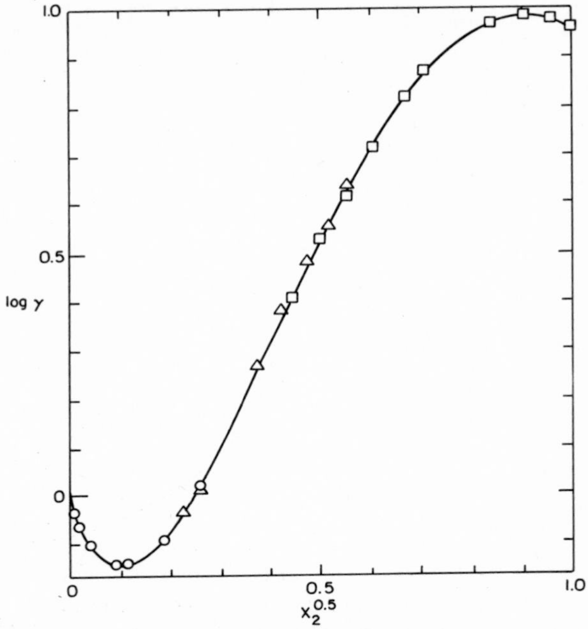

Fig. 8.6. Nitric Acid (25 C. o Landolt-Börnstein; □ Potier; Δ Haase)

redefined by means of the mole fraction x_2 according to

$$a_2 = (x_2\gamma)^2; \qquad \lim_{x_2=0} a_2 = x_2^2 \tag{8.92}$$

because the molality m tends to infinity at $x_2 = 1$ and is therefore unsuitable for high concentrations. In order to convert the usual activity coefficient γ^* defined by

$$a_2^* = (m\gamma^*)^2; \qquad \lim_{m=0} a_2 = m^2 \tag{8.93}$$

into γ we take into account that $\log a_2$ and $\log a_2^*$ are related to free energies of two standard states and therefore differ for any concentration by a constant amount

$$\log a_2 - \log a_2^* = \lim_{x_2=0} (\log a_2 - \log a_2^*)$$

$$= \lim_{x_2=0} [\log(x_2/m)^2] = \log(0.018016^2) \tag{8.94}$$

(cf. Eq.(7.48)). The conversion is given, according to the last three equations, by

$$\gamma = a_2^{0.5}/x_2; \qquad \gamma^* = a_2^{*0.5}/m; \qquad (a_2/a_2^*)^{0.5} = 0.018016 \tag{8.95}$$

$$\gamma/\gamma^* = 0.018016m/x_2 = 1 + 0.018016\ m. \tag{8.96}$$

The vapor pressures of nitric acid solutions are small enough to allow us to disregard deviations of the vapor from the perfect gas. The partial pressures of HNO_3 are therefore proportional to the activity so that we can compute the activity except for a constant factor.

Figure 8.6 shows log γ derived from freezing points. The results from vapor pressures have been shifted until the curve fitted smoothly to the freezing point results. The shift amounted to 0.957 in log γ or 1.914 in log a_2, representing the difference of the standard states for $x_2 = 0$ and $x_2 = 1$.

The activity coefficients of perchloric acid present a similar problem [85.4]. For sulfuric acid accurate measurements of electromotive forces supplement the other data so that the activity coefficients can be quite accurately derived over the whole concentration range [85.5].

8.6 DISSOCIATION

The quantitative representation of the dissociation of the weak electrolytes was the first major success of the theory of Arrhenius. At the same time (1888) its shortcomings in the question of strong electrolytes became manifest [86.1].

8.61 Thermodynamics of Dissociation. It will be sufficient to consider a binary dissociation, e.g., that of acetic acid. Abbreviating Ac for $CH_3 \cdot CO$ we have

$$\begin{array}{ccccc} \text{AcOH} & = & \text{AcO}^- & + & \text{H}^+ \\ (1-\alpha)m & & \alpha m & & \alpha m \end{array} \tag{8.97}$$

with m representing the stoichiometric (or analytical) molality and α the degree of dissociation.

The "true" molality of the ions is αm and their partial molal Gibbs free energy $\bar{G}_i$ can be represented by

$$\bar{G}_i = \bar{G}_i^\circ + 2\,RT\,\ell n(\alpha m \beta_i) \tag{8.98}$$

where $\bar{G}_i^\circ$ refers to a standard state in which both ions are present at unit activity. The "true" activity coefficient of the ions is now represented by β_1. In the range of validity of the theory of Debye and Hückel β_i is given by (8.43) with the modification that m is to be replaced by the "true" molality αm so that for a uni-univalent electrolyte (w=1)

$$\log \beta_i = -h(s^\circ \alpha m)^{0.5}. \tag{8.99}$$

The "true" molality of the undissociated molecules is $(1-\alpha)m$ and their partial molal free energy is represented by

$$\bar{G}_u = \bar{G}_u^\circ + RT\, \ell n[(1-\alpha)m\beta_u] \tag{8.100}$$

where the "true" activity coefficient β_u can be assumed to be unity over the range in which (8.99) is valid.

The equilibrium condition is now

$$\Delta G = \bar{G}_i - \bar{G}_u = \Delta G^\circ + RT\, \ell n[(\alpha m\beta_i)^2/(1-\alpha)m\beta_u] = 0. \tag{8.101}$$

Abbreviating

$$\ell n\, K = -\Delta G^\circ/RT \tag{8.102}$$

we have for the dissociation constant

$$K = \frac{\alpha^2\beta_i^2 m}{(1-\alpha)\beta_u}. \tag{8.103}$$

In Eq.(8.8) we introduced the partial molal free energy of a binary electrolyte as

$$\bar{G}_2 = \bar{G}_2^\circ + 2RT\, \ell n(m\gamma). \tag{8.104}$$

Except for the notation, this equation is identical with (8.98) since either represents the increase of free energy of a large amount of solution on addition of one mole of the electrolyte. Moreover, Eq.(8.100) represents the same quantity as long as there is equilibrium between ions and undissociated molecules (cf. Eq.(8.101)).

We conclude from the identity of (8.98) and (8.104) that the meaning of the stoichiometric activity coefficient is given by

$$\gamma = \alpha\beta_i. \tag{8.105}$$

The dissociation constant (8.103) has been derived and discussed by several authors [86.2] soon after Debye and Hückel developed their theory.

8.62 Weak Electrolytes. The dissociation constant of acetic acid at 25°C based on very accurate conductivity measurements [86.3] is shown in Fig. 8.7. The circles represent log K (Eq.(8.103)) calculated with the Debye approximation (8.99) for β_i and with $\beta_u = 1$. The abscissa is the square root of the ionic concentration αc (moles per liter). The constancy of log K at the level 0.2435-5 ($K = 1.753\cdot 10^{-5}$) is excellent up to $(\alpha c)^{0.5} = 0.02$ which corresponds to a stoichiometric concentration of about c = 0.01. The downtrend

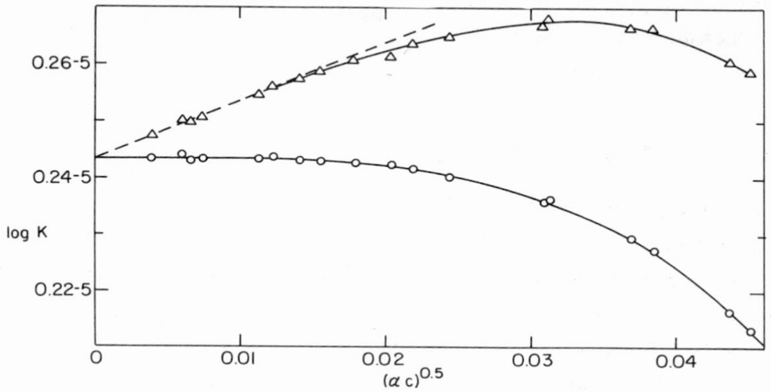

Fig. 8.7. Acetic Acid (25·C. o log K; Δ log K')

at higher concentrations illustrates the shortcoming of the approximation introduced for β_i and β_u. The results are in perfect agreement with measurements of the electromotive force [85.2]. The experimental methods will be briefly discussed in Section 8.64.

The triangles in Fig. 8.7 show the "classical" dissociation constant

$$K' = \alpha^2 c/(1-\alpha) \tag{8.106}$$

which is based on the assumption of a perfect solution, i.e., $\beta_i = 1$ and $\beta_u = 1$. The broken line indicates the slope $-2 \log \beta_i$ of the limiting law. Data of a very high accuracy such as these show of course the trend persisting to zero concentration. But on the whole the variation of K' is not large; the greater experimental errors of earlier times could completely conceal this trend. For this reason the interpretation of the dissociation of numerous weak electrolytes has rightly been considered a significant success of the classical theory of Arrhenius. Even for moderately strong electrolytes [86.4] with dissociation constants between 0.2 and 0.3 (trichloroacetic acid, iodic acid, thallous chloride) significant results, though within wide margins of error, were obtained by the classical methods.

The simple principle was to consider electrolyte solutions to be perfect, except of course for the dissociation. Quantitative calculations were therefore as convenient as calculations for perfect gaseous mixtures, and the existing analogy was often overdone. But there was a large, legitimate field of application to a variety of problems such as hydrolysis, galvanic cells, solubility, formation of complexes, and rate of reaction.

8.63 Acidity. For a long time acidity was considered to be synonymous with hydrogen ion concentration. Though Lewis [86.5] clearly divorced the two ideas by demonstrating the existence of proton-free acids, the "hydrogen ion concentration" c_{H^+} or its negative logarithm

$$pH = -\log c_{H^+} \tag{8.107}$$

has remained a practically important tool.

Conceptually the situation of pH is precarious. Though introduced by Sörensen with reference to a concentration, practically all measurements and applications aim at an activity. But this activity of a single ion has no thermodynamic basis. It is well-defined by virtue of the theory of Debye and Hückel for solutions of a low ionic strength because there we can reliably calculate the activity coefficient of any ion. But the term pH is almost always used for solutions whose ionic strength is far outside the range of the limiting law.

An escape has been found in conventional definitions based on certain galvanic cells, standard solutions for certain values of pH, and standardized colorimetric or spectrophotometric measurements with indicators (weak acids in which the ratio of anion to undissociated molecule depends on the hydrogen activity). Such conventional definitions are essentially arbitrary, but the various methods give similar results, though the deviations are often considerable. The importance of pH in technical operations, biology, organic chemistry and chemical kinetics has entailed a very great effort in the investigation of pH-measurement [86.6].

8.64 The Degree of Dissociation. The degree of dissociation α was introduced in Section 8.62 without discussion. The methods of its determination are at the core of the problem of the strong electrolytes.

The oldest method starts from the electrical conductivity. The molal conductivity Λ is the contribution per mole of an electrolyte to the transport of electric charge (through 1 cm^2 under a voltage gradient of 1 volt$\cdot$$cm^{-1}$). In the limit of infinite dilution the electrolyte is completely dissociated and the ions migrate without mutual interference; the extrapolated molal conductivity is denoted by Λ_0 and interpreted as the sum of the mobilities of the ions. If at any concentration only the fraction α of an electrolyte is dissociated, it contributes only $\alpha\Lambda_0$ to the molal conductivity, provided the motion of the ions is unimpeded. Then the degree of dissociation is given by

$$\alpha = \Lambda/\Lambda_0 \ . \tag{8.108}$$

This is the classical relation, which was so successful in the interpretation of weak electrolytes.

According to the conductivity theory developed by Debye and Onsager the interaction between ions reduces the mobilities at finite concentration. One can estimate the function Λ_e which replaces Λ_o in (8.108) from Kohlrausch's empirical law of independent migration: It says that the mobility of an ion depends on the concentrations and valence types of all ions present but not on the nature of the other ions. In other words, for four solutions of identical concentrations we have

$$\Lambda(HNO_3) + \Lambda(NaCl) = \Lambda(NaNO_3) + \Lambda(HCl). \qquad (8.109)$$

This rule is closely satisfied for strong electrolytes in dilute solutions.

It does not hold for weak electrolytes. But now we can estimate the sum of the mobilities Λ_e of acetic acid of concentration c by using the conductivities of three strong electrolytes at the concentration αc

$$\Lambda_e(AcOH) = \Lambda(NaOAc) + \Lambda(HCl) - \Lambda(NaCl). \qquad (8.110)$$

This procedure [86.3] represents a considerable improvement over (8.108).

The second method, extensively applied, is the interpretation of purely thermodynamic data, usually of Eq.(8.105). For weak electrolytes, the "true" activity coefficient is assumed to be unity; for dilute solutions of moderately strong electrolytes the limiting law gives a good refinement. The stretching or modification of the limiting law to cover higher concentrations inevitably introduces an element of serious uncertainty [86.7].

A few times information on degrees of dissociation has been derived from heat contents and molal volumes.

It can be shown [86.4] that all these methods fail for strong electrolytes, the upper limit being about $K = 0.5$ for uni-univalent electrolytes, lower for other types. Our knowledge of the "true" activity coefficients is not good enough to isolate the influence of incomplete dissociation for stronger electrolytes.

Optical methods have been often considered for the determination of the concentration of ions. It is reasonable to expect that light absorption is less influenced by peripheral ionic interaction than heat content or entropy. But a thorough analysis [86.7] does not quite confirm this expectation. Spectrophotometric investigation, however, allows a wide variety of applications and there is no general reason to exclude future success.

8.65 Strong Electrolytes. Two methods led finally to definite information of ion concentrations in solutions of strong electrolytes and also of their molecular structure.

Irradiation of a substance with light of a definite frequency often produces scattered light of different frequencies. The shift, discovered by Raman and by Landsberg and Mandelstam, is determined by vibration frequencies of the molecule and therefore a characteristic property of the molecule or ion. Moreover, in general neither frequency nor intensity of the Raman lines is very sensitive to the environment. The derivation of the degree of dissociation of nitric acid from the intensities of Raman lines of the nitrate ion was therefore repeatedly attempted. Definite quantitative results were first obtained by photographic comparison of the intensity of nitrate lines in solutions of nitric acid and of sodium nitrate [86.8]. A dissociation constant as high as $K = 21$ $(\text{mole}\cdot\text{lit}^{-1})$ was found. More accurate photoelectric measurements by Young [86.9] and his coworkers resulted in $K = 23.5$.

The derivation of K from observed degrees of dissociation is based on Eqs.(8.103) and (8.105).

Introducing the activity of nitric acid

$$a_2 = \gamma^2 m^2 \qquad (8.111)$$

and redefining β_u by referring it to c, we have

$$K = a_2/[(1-\alpha)c\beta_u]. \qquad (8.112)$$

We plot

$$\log(K\beta_u) = \log[a_2/(1-\alpha)c] \qquad (8.113)$$

against c and obtain log K by extrapolation to $c = 0$, which we choose as the standard state $\beta_u = 1$.

The second method is based on the fact that the nuclear magnetic resonance (NMR) frequencies of various atoms (H,N,Cl,F) are noticeably different [86.10] for various ions and molecules in which they are bound differently. The investigation of nitric acid [86.11] furnished an excellent confirmation of the Raman results.

In the gap between thallous chloride ($K = 0.3$) and nitric acid a few other electrolytes [86.12] are known, e.g., trifluoroacetic acid ($K = 1.8$) and heptafluorobutyric acid ($K = 1.1$). Still a much stronger electrolyte than nitric acid is perchloric acid [86.11].

As a combined result of investigations of various kinds the molecular states of strong acids can be described in some detail. Anhydrous perchloric acid appears to be monomeric. Water reacts with the acid to a solvated hydrogen ion and perchlorate ion. On further dilution the solvating acid molecules are gradually replaced by water molecules. The degree of dissociation is high even in the equimolal solution.

Nitric acid is more complicated. The anhydrous acid contains a small amount of the ions NO_2^+ and NO_3^-. The remainder is polymerized to dimers or even higher. On dilution the water is practically completely used up to convert the dimers into monohydrates. The degree of dissociation of the equimolal solution is small; it consits mainly of monohydrate. The dissociation progresses on further dilution and is practically complete at a concentration of about 4 moles per liter.

The state of sulfuric acid has been elucidated by Young and his coworkers [86.9].

The combination of activity coefficients and degrees of dissociation has quantitatively explained a good deal of the chemical behavior of these acids. Oxidation reactions and explosions are linked to the undissociated molecules.

SUMMARY OF THE EIGHTH CHAPTER

The unique importance of water and electrolytes in our lives follows from the fact that electrostatic forces in water are eighty times smaller than in the vacuum. There are very few solvents with a dielectric constant as high. For this reason electrostatic attraction prevents extensive dissociation in most non-aqueous solutions.

The decisive difference between ordinary molecular interaction and ionic interaction is caused by the range of the forces involved. Intermolecular forces have been well represented as functions of the distance r by terms proportional to r^{-6} or higher. Coulomb forces, however, decrease proportionally to r^{-2}; their influence, therefore, extends over much longer distances. It is this feature of electrostatic interaction which causes the characteristic terms with $c^{0.5}$ predicted by Debye and Hückel in the free energy and other thermodynamic functions. Square-root terms have an appreciable influence at much lower concentrations (higher interionic distances) than first-order terms, which are the rule for the first derivations of non-electrolyte solutions from the ideal law.

Electrolytes and their properties constitute by themselves an important part of chemical thermodynamics. Moreover, they have provided methods for thermodynamic measurements and for the investigation of the molecular structure of solutions.

References

(82.1) G. N. Lewis and M. Randall, J. Am. Chem. Soc., 43 (1921) 1112.

(83.1) R. Malmström, Z. Elektrochem., 11 (1905) 797.

(83.2) P. Debye and E. Hückel, Physik. Z., 24 (1923) 185.

(83.3) M. von Laue, Jahrbuch Radioakt., 15 (1918) 205.

(83.4) P. Gross and O. Halpern, Physik. Z,, 26 (1925) 403.

(83.5) N. Bjerrum, Z. physik. Chem., 119 (1926) 145.

(83.6) O. Redlich and P. Rosenfeld, Z. physik. Chem., A 155 (1931) 61. Z. Elektrochem., 37 (1931) 705.

(83.7) W. Geffcken et al., Z. physik. Chem., B 20 (1933) 398. 23 (1933) 175. 35 (1935) 317. O. Redlich and J. Bigeleisen, J. Amer. Chem. Soc., 64 (1942) 758. Chem. Revs., 30 (1942) 171.

(83.8) L. A. Dunn, Trans. Faraday Soc., 62 (1966) 2348. 64 (1968) 1898, 2951. F. J. Millero, J. Phys. Chem., 71 (1967) 4567. 74 (1970) 356.

(83.9) T. F. Young, Y. C. Wu and A. A. Krawetz, Discussions Faraday Soc. 24 (1957) 37, 77, 80. Y. C. Wu, M. B. Smith and T. F. Young, J. Phys. Chem., 69 (1965) 1873.

(84.1) W. M. Latimer, The Oxidation States of the Elements and Their Potentials in Aqueous Solutions, Prentice-Hall, New York, 1938, 1952.

(85.1) Lewis-Randall-Pitzer-Brewer, Thermodynamics, 2nd ed. McGraw-Hill, New York, 1961. Chapter 26.

(85.2) E. Abel, O. Redlich and B. Lengyel, Z. physik. Chem., 132 (1928) 189. F. Hartmann and P. Rosenfeld, ibid. A 164 (1933) 377. Calculations: Landolt-Börnstein, Tables, 5th Edition, 3rd Suppl. p. 2145, Springer, Berlin, 1935.

(85.3) A. Potier, Ann. Fac. Sci. Univ. Toulouse, 20 (1956) 1. R. Haase K. H. Dücker and H. A. Küppers, Ber. Bunsen-Gesellschaft, 69 (1965) 97.

(85.4) O. Redlich, W. E. Gargrave and W. D. Krostek, I&EC Fundamentals, 7 (1968) 211.

(85.5) J. I. Gmitro and T. Vermeulen, AI.Ch.E. Journal, 10 (1964) 740.

(86.1) S. Arrhenius, J. Amer. Chem. Soc., 34 (1912) 353. A vivid description of the early history of the theory.

(86.2) P. Gross and O. Halpern, Physik. Z., 25 (1924) 393. M. S. Sherrill and A. A. Noyes, J. Amer. Chem. Soc., 48 (1926) 1861. D. A. MacInnes, J. Amer. Chem. Soc., 48 (1926) 2068.

(86.3) D. A. MacInnes and T. Shedlovsky, J. Amer. Chem. Soc., 54 (1932) 1429.

(86.4) O. Redlich, Chem. Revs., 39 (1946) 333. Survey.

(86.5) G. N. Lewis, J. Franklin Inst., 226 (1938) 293.

(86.6) R. G. Bates, Determination of pH, Theory and Practice, Wiley, New York, 1964.

(86.7) J. E. Prue, Discussions Faraday Soc., 24 (1957) 115. W. G. Davies, R. J. Otter and J. E. Prue, ibid. 103. E. A. Guggenheim, ibid. 81.

(86.8) O. Redlich and P. Rosenfeld, Monatsh., 67 (1936) 223. O. Redlich and J. Bigeleisen, J. Amer. Chem. Soc., 65 (1943) 1883. N. R. Rao, Indian J. Physics, 15 (1941) 185.

(86.9) T. F. Young, Record Chem. Progr. (Kresge-Hooker Sci. Lib.), 12 (1951) 81. T. F. Young, L. F. Maranville and H. M. Smith in "The Structure of Electrolytic Solutions," W. J. Hamer, Ed. (John Wiley, New York, 1959), p. 35.

(86.10) H. S. Gutowsky and A. Saika, J. Chem. Phys., 21 (1953) 1688. Y. Masuda and T. Kanda, J. Phys. Soc. Japan, 8 (1953) 432.

(86.11) G. C. Hood, O. Redlich and C. A. Reilly, J. Chem. Phys., 22 (1954) 2067. O. Redlich, R. W. Duerst and A. Merbach, ibid. 49 (1968) 2986.

(86.12) O. Redlich and G. C. Hood, Discussions Faraday Soc., 24 (1957) 87.

CHAPTER 9: CHEMICAL REACTIONS

The preceding chapters have prepared, one by one, the tools we need in the thermodynamic examination of chemical reactions. The remaining task is the application of these tools and a discussion of the methods used in drawing conclusions from available data and predicting reaction results. This discussion will lead only to a few general relations and will be carried mostly by illustrative examples.

9.1 THE EQUILIBRIUM CONSTANT

The usefulness of chemical thermodynamics stems from the fact that calorimetric measurements as a rule can be carried out much more easily than quantitative chemical determinations. For this reason one attemps to squeeze out as much information from thermal data as possible. The basis for chemical applications is furnished by Eq.(5.14)

$$\Delta G = \Sigma \nu_j \bar{G}_j = 0 \tag{9.1}$$

for the change of the Gibbs free energy in a process in which no useful work is done. As a rule, the symbol Δ is used for a reaction specified at a certain temperature and pressure by the reaction coefficients ν_i. If, as usual, temperature, pressure and numbers of moles are chosen as independent variables, the order of the symbols Δ, $\partial/\partial T$ and $\partial/\partial P$ can be arbitrarily changed.

The usual and convenient procedure of application rests on the equilibrium constant. An example has been anticipated in the dissociation constant of Eq.(8.103). The equilibrium constant is used to describe the influence of the concentrations of solutions on the equilibrium at a given temperature and pressure. In all discussions of gases and solutions we split off a standard term $\bar{G}_j^o$ (independent of pressure and concentration) from the partial molal Gibbs free energy $\bar{G}_j$. The remaining terms contain either (Eqs.(3.60) and (6.13)) the fugacity $Py_j\phi_j$ or (Eqs.(6.30) and (8.7)) the activity

$$a_j = x_j\gamma_j \quad \text{or} \quad a_j = (m\gamma)^\nu \tag{9.2}$$

(where ν is the split factor of an electrolyte). Generally we may say

$$\bar{G}_j = \bar{G}_j^o + RT \ln a_j \tag{9.3}$$

if we keep in mind that for gases we have to replace a_j by the fugacity and that for liquid or solid substances we may have to add a Poynting term (Eq.(6.33)).

Then we obtain the equilibrium condition

$$\Delta G = \Sigma\nu_j\bar{G}_j = \Sigma\nu_j(\bar{G}_j^o + RT\,\ell n\,a_j) = \Delta G^o + RT\Sigma\nu_j\,\ell n\,a_j = 0. \quad (9.4)$$

If we now introduce the equilibrium constant K by means of

$$\ell n\,K = -\Delta G^o/RT = -\Sigma\nu_i\bar{G}_i^o/RT \quad (9.5)$$

the equilibrium condition can be written

$$\Sigma\nu_i\,\ell n\,a_i = \ell n\,K. \quad (9.6)$$

In this relation a_i may be represented by means of pressure, mole fraction and fugacity coefficient, or by means of mole fraction and activity coefficient, or by molality and mean activity coefficient, or otherwise. But the choice is tied to the choice of the standard state and the value of $\bar{G}_j^o$.

The equilibrium constant is a function of the temperature but independent of the pressure. From (3.8) and (9.5) we obtain

$$\frac{d\,\ell n\,K}{dT} = -\frac{1}{R}\cdot\frac{d(\Delta G^o/T)}{dT} = \frac{\Delta H^o}{RT^2}. \quad (9.7)$$

The meaning of ΔH^o is the change in heat content of the reaction proceeding from standard state to standard state, or the (negative) heat of reaction. Its temperature dependence is given by (3.7) as

$$\left(\frac{\partial\Delta H^o}{\partial T}\right)_P = \Delta C_p^o. \quad (9.8)$$

Often this quantity is small so that

$$\frac{d\,\ell n\,K}{d(1/T)} = -T^2\frac{d\,\ell n\,K}{dT} = -\frac{\Delta H^o}{R} \quad (9.9)$$

is nearly independent of the temperature. In a diagram with 1/T as the abscissa, one obtains a straight line or a slightly curved line for ℓn K.

Since $\bar{G}_j^o$ is <u>defined</u> as independent of pressure, we conclude from (9.5) for reactions in the gas phase

$$\frac{\partial\,\ell n\,K}{\partial P} = -\Sigma\nu_j\frac{\partial\,\ell n(Py_j\phi_j)}{\partial P}$$

$$= -\frac{\Sigma\nu_j}{P} - \frac{\Sigma\nu_j\partial\ell n(y_j\phi_j)}{\partial P} = 0. \quad (9.10)$$

If we write

$$\ell n\,K_y = \Sigma\nu_j\,\ell n\,y_j \quad (9.11)$$

we have for a reaction within a <u>perfect</u> gas mixture (with $\bar{V}$ for the molal volume and $\phi_j = 1$)

$$\frac{\partial \ln K_y}{\partial P} = -\frac{\Sigma\nu_j}{P} = -\frac{\bar{V}\Sigma\nu_j}{RT} = -\frac{\Delta V}{RT} . \qquad (9.12)$$

But if we introduce (for a perfect gas mixture) the partial pressures $p_j = Py_j$ the resulting equilibrium constant is still K and therefore independent of the pressure.

For a liquid or solid solution we have to add Poynting terms in (9.3) to (9.5) so that differentiation of (9.5) furnishes

$$RT\ \partial \ln K/\partial P = -\Sigma\nu_j\bar{V}_j^o = -\Delta V^o , \qquad (9.13)$$

in general only a small quantity for liquids and solids.

In the following sections we wish to illustrate the principles of applying chemical thermodynamics rather than the technique. Details of computation methods will be discussed in Appendix 12.

Modern chemical technology has been the answer to a demand for large amounts of cheap materials such as fertilizers. The control of gaseous reactions on a large scale under conditions selected by painstaking research has been the first step. Haber's ammonia synthesis is an example of the many and great benefits that Western civilization has bestowed on the whole of mankind: A large and increasing part of the food of the world depends on this process, the work of a few great men and of innumerable devoted scientists and engineers.

9.2 AMMONIA SYNTHESIS

The hot problem of the first decade of this century was "nitrogen fixation", the conversion of atmospheric nitrogen to compounds that are readily assimilated by plants. At that time available natural fertilizers became insufficient while population was increasing rapidly. The direct synthesis of ammonia at elevated temperature and pressure turned out to be far superior to various other attempts (heating of air, usually in the electric arc, to 2000°C and rapid cooling to conserve the small amount of nitric oxide formed, production of calcium cyanamide from carbide and nitrogen). Synthetic ammonia indeed has become the source of all man-made nitrogen compounds. It is an indispensable protection against wide-spread starvation.

The path to a technical process was beset with numerous difficulties. At room temperature nitrogen and hydrogen do not react. The reaction rate increases with increasing temperature but at the same time the equilibrium becomes less favorable for the formation of ammonia. Higher pressure improves the ammonia yield but the handling of large amounts of gases at high temperature and pressure presented serious problems. Recycling of the unconverted nitrogen-hydrogen mixture was obviously necessary but argon introduced with atmospheric nitrogen accumulated in the recycled gas and had to be bled off with inevitable losses of feed. Another big problem was the development of an efficient catalyst and its protection from poisons such as hydrogen cyanide and carbon monoxide.

There are answers to every one of these problems. But they are all interconnected and the remedies must be balanced. Argon, for instance, can be removed from nitrogen by distillation at low temperature, a fairly expensive step; but then one can use a lower pressure because the lower bleeding requirement makes an increase of the recycle acceptable.

The contribution of thermodynamic calculations to the solution of these problems of design balancing is the complete description of the possible equilibrium yield as a function of temperature, pressure and feed composition. The calculation requires three steps:

(a) The computation of the equilibrium constant from available free energy data.
(b) The estimation of the fugacity coefficients.
(c) The computation of the yields for suitably selected temperatures, pressures and feed compositions.

The first step is simple if we consult the tables of Wicks and Block [91]. Here we find a formula for ΔG° (denoted by the authors ΔF_T) for the ammonia synthesis

$$\underset{(1)}{0.5\ N_2} + \underset{(2)}{1.5\ H_2} = \underset{(3)}{NH_3} \qquad (9.14)$$

and also a table from which we take

$$\Delta G^\circ = 3700 \text{ cal } (600\ K); \qquad \Delta G^\circ = 9200 \text{ cal } (800\ K). \qquad (9.15)$$

For the following computation we interpolate for 450°C

$$\Delta G^\circ = 6986\ (723.15\ K) \qquad (9.16)$$

so that according to (9.5)

$$\log K = -6986/(2.3026 \cdot 1.9872 \cdot 723.15) = -2.1113. \qquad (9.17)$$

We split

$$\log K = \log(P_{y_3}\phi_3/[(P_{y_1}\phi_1)^{0.5}(P_{y_2}\phi_2)^{1.5}]) \qquad (9.18)$$

according to

$$\log K = \log K_y + \log K_\phi - \log P \tag{9.19}$$

$$K_y = y_3/(y_1^{0.5}y_2^{1.5}) \tag{9.20}$$

$$K_\phi = \phi_3/(\phi_1^{0.5}\phi_2^{1.5}) \; . \tag{9.21}$$

The temperature of 450 C is so high above the critical temperatures of the three substances involved that the equation of state (3.26) of Redlich and Kwong may be expected to furnish a good estimate of the fugacity coefficients even at high pressures. We use relation (6.28) to compute ϕ_1, ϕ_2, ϕ_3 and K_ϕ for various values of y_3. The results are shown in Figs. 9.1 and 9.2.

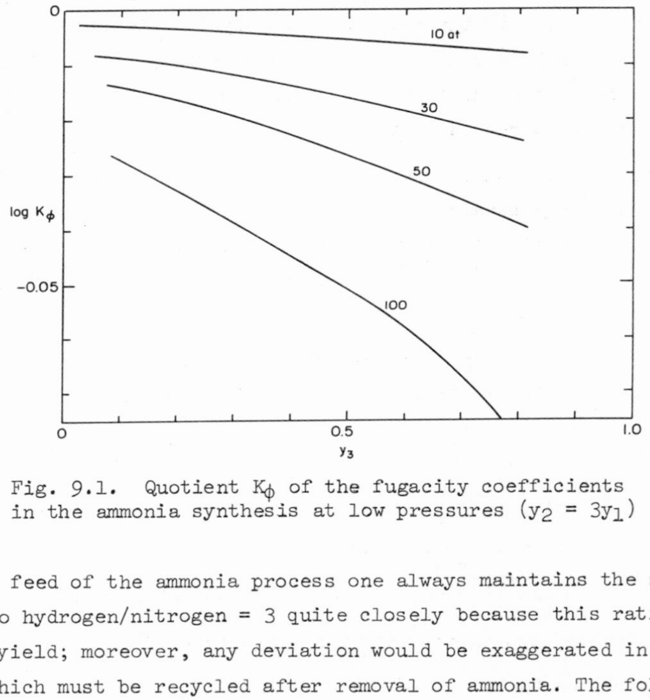

Fig. 9.1. Quotient K_ϕ of the fugacity coefficients in the ammonia synthesis at low pressures ($y_2 = 3y_1$)

In the feed of the ammonia process one always maintains the stoichiometric ratio hydrogen/nitrogen = 3 quite closely because this ratio furnishes an optimum yield; moreover, any deviation would be exaggerated in the reactor effluent, which must be recycled after removal of ammonia. The following calculation is therefore restricted to a stoichiometric feed. The quotient of fugacity coefficients K_ϕ has been computed for mixtures as a function of y_3 with the stoichiometric ratio $y_2/y_1 = 3$, so that $y_1 = 0.25(1-y_3)$ and $y_2 = 0.75(1-y_3)$. Table 9.1 illustrates the representation of the equilibrium mole fractions y_1, y_2, y_3 by the conversion c, i.e., the fraction of feed converted to ammonia.

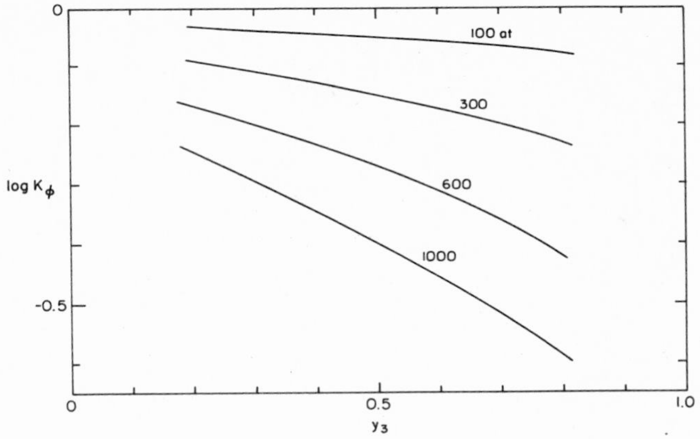

Fig. 9.2. Quotient K_ϕ at high pressures

Table 9.1

Numbers of Moles in the Ammonia Process

	N_2 (1)	H_2 (2)	NH_3 (3)	Σ
Feed	0.5	1.5	0	2
Equilibrium mixture	0.5(1-c)	1.5(1-c)	c	2-c
Mole fractions	0.5(1-c)/(2-c)	1.5(1-c)/(2-c)	c/(2-c)	1

The equilibrium constant (Eq.(9.17)) and the quotient K_ϕ furnish the data for the computation of the mole fraction y_3 of ammonia. We obtain from (9.19), (9.20) and Table 9.1

$$K_y = PK/K_\phi = \frac{c}{2-c}\left(\frac{2-c}{0.5(1-c)}\right)^{0.5}\left(\frac{2-c}{1.5(1-c)}\right)^{1.5} \tag{9.22}$$

$$\rho = 0.5^{0.5} \cdot 1.5^{1.5}\ PK/K_\phi = c(2-c)/(1-c)^2 \tag{9.23}$$

$$c = 1 - 1/(\rho+1)^{0.5} \tag{9.24}$$

The conversion c (which must be smaller than unity) is given by the quadratic equation (9.23), which is solved in (9.24). The actual computation is completely recorded in Table 9.2. No notes need be taken outside this simple and brief tabular computation sheet for completion with the aid of a desk calculator or a slide rule.

Table 9.2

Equilibrium in the Ammonia Process

[450°C; log K = -2.1113; $\log(0.5^{0.5} \cdot 1.5^{1.5} K) = -1.9977$]

P(atm)	10	30	50	100	300	600	1000
y_3(est.)	.02	.06	.09	.16	.36	.56	.70
log K_ϕ	-.0025	-.0085	-.0135	-.0305	-.120	-.282	-.517
log P	1.0000	1.4771	1.6990	2.0000	2.477	2.778	3.000
log ρ	-.8989	-.5121	-.2852	.0328	.599	1.062	1.519
ρ	.1024	.3075	.5182	1.077	3.972	11.51	33.11
$(\rho+1)^{-0.5}$	.954	.875	.811	.693	.449	.294	.171
c	.046	.125	.189	.307	.551	.706	.829
y_3(calc.)	.0245	.0666	.1041	.181	.381	.545	.706
y_3(obs.)	.0211	.0586	.0915	.1643	.3582	.5371	.6969

Estimated values of y_3 are required for finding the values of K_ϕ from the diagrams. If the estimate turns out to be wrong, one has to repeat the computation with a better estimate. More than two repetitions are rarely necessary.

The agreement of the calculated and observed [92] results is surprisingly good, particularly in view of the prediction of K_ϕ by an equation based only on the critical temperatures and pressures of the substances involved.

The results shown in Table 9.2 illustrate the important influence of the pressure in the ammonia synthesis.

9.3 SULFURIC ACID

A large part of the chemical industry depends on sulfuric acid. The core of its production is the catalytic oxidation of sulfur dioxide (contact process)

$$\underset{(1)}{SO_2} + \underset{(2)}{0.5\, O_2} = \underset{(3)}{SO_3} \tag{9.25}$$

developed by Knietsch.

The problem of choosing the temperature is similar to that of the ammonia process: With increase of temperature the equilibrium shifts to increasing residues of sulfur dioxide, and with decreasing temperature the reaction rate (with both the early catalyst platinum and the present vanadium pentoxide) drops. The reaction is carried out in several steps starting at higher temperatures and ending at about 450°C.

Increase of pressure offers less advantage than in the ammonia synthesis since the relative volume reduction is smaller. Moreover, since the feed gas is obtained by combustion of sulfur or roasting of sulfides, it always contains a large amount of nitrogen which would greatly increase the cost of compression. Atmospheric pressure, therefore, is chosen and deviations from the perfect gas are negligible at the high temperature of the reaction.

Some excess of oxygen must be introduced for the sufficient utilization of sulfur. The degree of conversion, or better, the final residue of sulfur dioxide is important not because of its value but because of pollution. The following example will therefore discuss the residue r (in moles SO_2/gram-atom S introduced or in t SO_2/1000 t H_2SO_4 produced) as a function of the excess ratio η of oxygen (ratio of O_2 introduced/O_2 stoichiometrically required). The example will be restricted to elemental sulfur as raw material. The calculation will be based on 1 gram atom sulfur so that 1.5 moles O_2 will be stoichiometrically required for the production of 1 mole H_2SO_4.

The thermodynamic information can be obtained from Appendix 7 of the "Thermodynamics" of Lewis, Randall, Pitzer and Brewer. The tables presented by Pitzer and Brewer are not as convenient as those of Wicks and Block but they contain a very large amount of information in a small space. These tables contain the quotients

$$q_o = -(G^\circ - H^\circ_o)/T \tag{9.26}$$

for a great number of elements and compounds between 298 and 2000 K. The Gibbs free energy G° (called F° in the tables) is the value for the formation of the compound at the given temperature in its standard state from the elements in their standard states. The symbol H°_o represents the integration constant involved in energy and free energy; it is chosen equal to zero for elements in their standard states at 0 K. For any compound i the heat content of formation at 0 K from the elements j is

$$H^f_{oi} = \Sigma\nu_j H^\circ_{oj} \tag{9.27}$$

The quantity H^f_{oi} is given in the tables under the name of ΔH°_o. The increase in heat content during reaction (9.25) at 0 K is now easily computed as

$$\Delta H^f_o = \Sigma\nu_i H^f_{oi}\ . \tag{9.28}$$

Applying (9.26) to the reaction (9.25) we have for the temperature T

$$\Delta q_o = \Sigma\nu_i q_{oi} = -\Delta G^\circ/T + \Delta H^f_o/T \tag{9.29}$$

so that

$$-\Delta G^\circ/T = \Delta q_o - \Delta H_o^f/T \tag{9.30}$$

$$\log K = -\Delta G^\circ/(2.3026\ RT)$$
$$= \Delta q_o/4.5756 - \Delta H_o^f/(4.5756\ T). \tag{9.31}$$

The computation of K is outlined in Table 9.3 for 450°C = 723.15 K.

Table 9.3

Equilibrium Constant for the Knietsch Process

(i)		SO_2 + (1)	$0.5\ O_2$ = (2)	SO_3 (3)	Δ
ν_i		-1	-0.5	1	
$q_o = -(G^\circ - H_o^\circ)/T$	500 K	55.38	45.68	57.14	-21.08
	1000 K	62.28	50.70	66.08	-21.55
	723 K				-21.29
H_o^f (kcal)		-70.36	0	-93.06	-22.70
$\Delta H_o^f/723.15$ (cal·deg^{-1})					-31.39
$\Delta G^\circ/723.15$					-10.10
log K					2.207
K					161.1

Since the gaseous mixture may be considered to be perfect, the equilibrium constant is

$$K = \frac{y_3}{y_1(y_2P)^{0.5}} = 161.1 \text{ (at 723.15 K)} \tag{9.32}$$

with P = 1 atm. It will enable us to compute the residue of SO_2 for a start from 1 gram-atom S and 1.5 η moles O_2.

The stoichiometry is shown in Table 9.4. Using air for oxidation, we

Table 9.4

Numbers of Moles in the Knietsch Process

	SO_2	O_2	SO_3	N_2	Σ
Feed (after combustion)	1	1.5η-1	0	5.625η	7.125η
Equ. mixture	r	1.5η-1.5+0.5r	1-r	5.625η	7.125η-0.5(1-r)
Approx. equ. mixture	r	1.5η-1.5	1	5.625η	7.125η-0.5
Stack gas	r	1.5η-1.5	0	5.625η	7.125η-1.5
Second equilibrium	rs	1.5η-1.5	r(1-s)	5.625η	7.125η-1.5

introduce 0.79/0.21 moles N_2 with each mole O_2 so that the amount of N_2 (including argon) in all steps is 5.625η. The combustion of sulfur to sulfur dioxide proceeds completely. The oxidation of 1-r moles SO_2 furnishes the amounts given in the second line of the table. With the numbers of moles given there we obtain the equilibrium condition

$$K = \frac{1-r}{r} \cdot \left(\frac{7.125\eta - 0.5 + 0.5r}{1.5\eta - 1.5 + 0.5r}\right)^{0.5} . \qquad (9.33)$$

This is a quadratic equation in η so that η can be computed directly for a series of values of r. If we have to prepare a computer program for routine work, we should solve this equation. Otherwise we better make use of the fact that in all interesting cases the residue r is small compared with unity. Since oxygen is always provided in excess and η is noticeably greater than unity, the residue r can be disregarded in the amounts of the reactants (second line in Table 9.4) except SO_2; thus one obtains the approximate amounts in the third line. Similarly (9.33) can be simplified so that r can be computed for any value of η by means of

$$r = \frac{1}{K} \cdot \left(\frac{7.125\eta-0.5}{1.5\eta-1.5}\right)^{0.5}$$

$$= 6.20 \cdot 10^{-3} \left(\frac{4.75\eta-0.333}{\eta - 1}\right)^{0.5} . \qquad (723.15\ K) \qquad (9.34)$$

The computation is outlined in Table 9.5. The mole fraction y_{1s} of SO_2 in the stack gas is given by

$$y_{1s} = r/(n_e - 1) \qquad (9.35)$$

since 1 mole H_2SO_4 is removed from the equilibrium mixture before it goes to the stack.

Table 9.5

Stack Gas in the Knietsch Process

Oxygen excess ratio	1.1	1.5	2.0	3.0	∞
Total mole number n_e in equil. (for 1 mole S)	7.29	10.17	13.7	20.9	∞
Residue r of SO_2 (mole/mole S)	0.0436	0.0234	0.0188	0.0164	0.0136
Mole fraction SO_2 in effluent	0.0070	0.0025	0.0015	0.0008	0
Pollution (t SO_2/1000t H_2SO_4)	29.8	15.6	12.5	10.9	8.95
After second equilibrium					
Ratio s of residues	0.0403	0.0216	0.0181	0.0159	0.0136
Residue rs	0.00176	0.00051	0.00034	0.00026	0.00019
Pollution (t/1000t)	1.15	0.33	0.22	0.13	0.13

A highly reputed construction company claimed in an advertisement several years ago that their sulfuric acid plants release not more than 0.2% SO_2 in the stack gas. Table 9.5 (line 4) shows that this claim is very reasonable and requires an oxygen excess ratio of slightly less than 2.0.

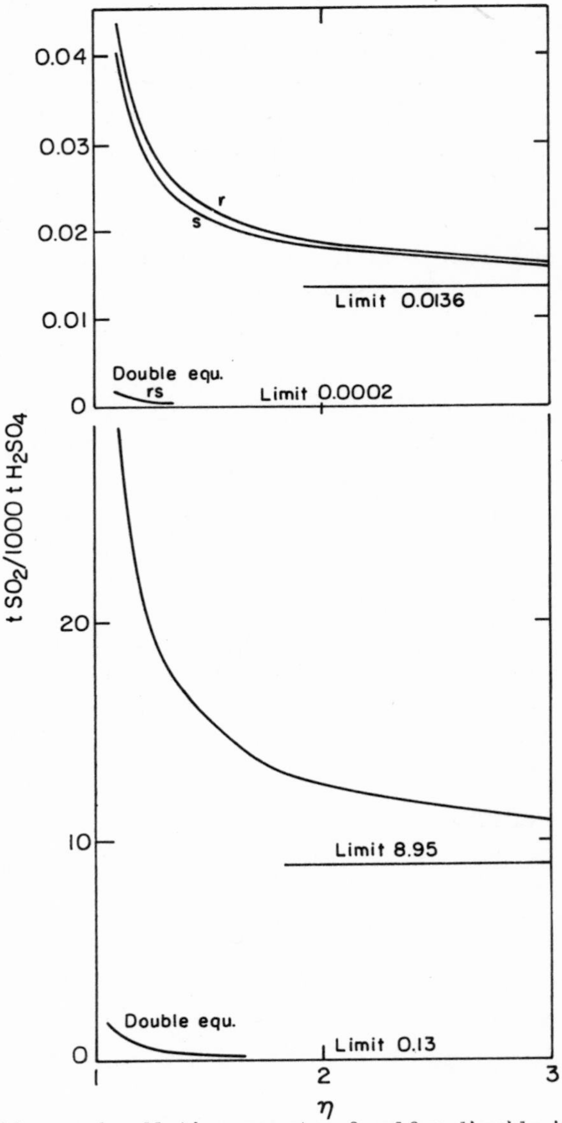

Fig. 9.3. Residues and polluting amounts of sulfur dioxide in the Knietsch Process (r residue after first equilibrium rs after the second)

But the computation shows one very disquieting result. We can easily compute the amount of SO_2 (in tons) evolved in the manufacture of 1000 t H_2SO_4; it is $(64/98)r/(1-r)$. The fifth line of Table 9.5 shows that this polluting amount is of the order of 10 t and cannot drop below 8.95 t even with increasing oxygen ratio.

A recently suggested modification [93] greatly reduces the amount of pollution. After absorption of H_2SO_4 the gas is reheated to 450°C and streams again over a catalyst. The computation starts now from r moles of SO_2 and proceeds in a similar way as before. We obtain the ratio s of the residues after and before the second equilibrium. The final residue rs and the final pollution values in Table 9.5 are encouragingly low, even if practical results must be expected to be somewhat inferior to the calculated equilibrium results. The results obtained for the residues r and rs and for the amounts of SO_2 in the stack gas are shown in Fig. 9.3.

The approximation method used here can be generally applied. We estimate at first equilibrium conditions disregarding impurities or other minor products. The results for the major constituents are then used for an estimate of the minor products.

9.4 CARBON DEPOSITION

Hydrogen has a higher value as a raw material for syntheses than as a fuel. The production of hydrogen by electrolysis of water was economically justified where there was a surplus of hydroelectric power. Today most of the hydrogen required is obtained by processes based either on coal or hydrocarbons from petroleum or natural gas. In these processes deposition of carbon presents a problem.

9.41 Outline of the Problem. Methane, the main constituent of natural gas, furnishes hydrogen on heating with steam by the reactions

$$\underset{(1)}{CH_4} + \underset{(2)}{H_2O} = \underset{(3)}{3H_2} + \underset{(4)}{CO} \tag{9.36}$$

$$CH_4 + 2H_2O = 4H_2 + \underset{(5)}{CO_2}. \tag{9.37}$$

The plain cracking reaction

$$CH_4 = \underset{(6)}{C(s)} + 2H_2 \tag{9.38}$$

also furnishes hydrogen but it is undesirable (except in the production of carbon black) because deposit of carbon would impede the gas flow through catalysts and pipes.

At the conditions of hydrogen production these three reactions give a practically complete description of the equilibrium. We may combine the equations in various ways to obtain other reactions, e.g., from the first two we find the so-called shift reaction

$$CO + H_2O \;=\; H_2 + CO_2 \tag{9.39}$$

or we may represent carbon deposition by

$$2CO \;=\; C(s) + CO_2 \tag{9.40}$$

(by adding Eq.(9.37) and (9.38) and subtracting twice Eq.(9.36)). Not more than three of these equations are independent and required for the determination of the equilibrium.

The thermodynamic analysis [94] is helpful because it outlines the conditions for attaining the main objectives of the process. These are: (a) High yield of hydrogen for one mole of methane. (b) High conversion of the feed methane to carbon dioxide rather than monoxide because monoxide is a catalyst poison and its removal more expensive. (c) No deposition of carbon.

The last requirement prescribes for the amount of carbon in equilibrium

$$n_6 \;=\; 0. \tag{9.41}$$

We feed (1-w) moles of methane and w moles of water and want to know the composition of gas mixture obtained at equilibrium at some temperature T and atmospheric pressure. Since the reactions (9.38) or (9.40) do not contribute to the process, the stoichiometric relations are completely prescribed by (9.36) and (9.37). We express the numbers of moles in equilibrium $n_1(CH_4)$, $n_2(H_2O)$, $n_3(H_2)$ by means of $n_4(CO)$ and $n_5(CO_2)$ and the feed composition

$$n_1 \;=\; 1 - w - n_4 - n_5 \tag{9.42}$$

because one mole CH_4 disappears for each mole CO and CO_2, and similarly

$$n_2 \;=\; w - n_4 - 2n_5 \tag{9.43}$$

$$n_3 \;=\; 3n_4 + 4n_5\ . \tag{9.44}$$

The total number of moles is

$$n \;=\; n_1 + n_2 + n_3 + n_4 + n_5 \;=\; 1 + 2n_4 + 2n_5\ . \tag{9.45}$$

and the mole fractions are

$$y_1 \;=\; n_1/n \text{ and so on.} \tag{9.46}$$

At 1 atm and the high temperatures of interest the mixture is a perfect gas.

The absence of carbon requires, according to (9.40), that

$$y_5/y_4^2 = n_5n/n_4^2 \geq K_D \tag{9.47}$$

if we denote by K_D the equilibrium constant for the deposition reaction (9.40). If this condition is not satisfied the reaction may proceed and thus produce carbon. We shall use (9.47) with the equality sign and therefore obtain the border line between deposition and non-deposition.

The problem is now to determine for a given temperature (and atmospheric pressure) and six unknowns w, n_1, n_2, n_3, n_4, n_5 or w, n, y_1, y_2, y_3, y_4 from the three stoichiometric equations (9.42), (9.43), (9.44) and three equilibrium conditions, namely, (9.47) and

$$y_3^3y_4/(y_1y_2) = n_3^3n_4/(n_1n_2n^2) = K_1 \tag{9.48}$$

$$y_3^4y_5/(y_1y_2^2) = n_3^4n_5/(n_1n_2^2n^2) = K_2 \tag{9.49}$$

for the reactions (9.36) and (9.37), or from equivalent relations. A variety of methods can be used to solve problems of this kind. Some of them will be discussed in the following. The availability of an automatic computer will favor analytical procedures over graphical computation.

The equilibrium constants K_1, K_2, K_D are supposed to be known.

9.42 Substitution. The old method of using the available equations for eliminating the unknowns one by one can be quite efficiently used in this problem. We shall express y_5, y_3, y_2, y_1 by y_4. We obtain from (9.47)

$$y_5 = K_Dy_4^2 , \tag{9.50}$$

from (9.44) and (9.46)

$$y_3 = 3y_4(1 + 1.333\ K_Dy_4) = 3y_4(1 + Cy_4). \tag{9.51}$$

The division of (9.48) by (9.49) furnishes

$$y_2 = (3\ K_DK_1/K_2)y_4^2(1+1.333\ K_Dy_4) = By_4^2(1+Cy_4) . \tag{9.52}$$

Similarly the square of (9.49) divided by (9.48) leads to

$$y_1 = (9K_2/K_1^2K_D)y_4^2(1+1.333\ K_Dy_4)^2 = Ay_4^2(1+Cy_4)^2 . \tag{9.53}$$

We introduce these four relations into

$$y_1 + y_2 + y_3 + y_4 + y_5 = 1 \tag{9.54}$$

and obtain

$$y_4 = 0.25 - y_4^2[5K_3 + B(1+Cy_4) + A(1+Cy_4)^2] \tag{9.55}$$

with the abbreviations

$$A = 9K_2/(K_1^2K_D); \qquad B = 3K_1K_D/K_2; \qquad C = 1.333\ K_D. \tag{9.56}$$

This equation shows immediately that y_4 must lie between 0 and 0.25 since the content of the brackets is necessarily positive. The equation is written in a form suitable for a step-by-step approximation if y_4 is close to 0.25 since the content of the brackets must be small in this case. Table 9.6 shows that this condition is satisfied for high temperatures, but not below 1000 K. For lower temperatures one will therefore turn (9.55) around and solve for y_4^4 rather than y_4. The result of ordering and rearranging the powers of y_4 is

$$y_4^4 = [1 - 4y_4 - (5K_D+A+B)y_4^2 - (2A+B)y^3]/(AC^2)\ . \tag{9.57}$$

Table 9.6

Equilibria of the Methane-Steam Reactions

T(K)	K_1	K_2	K_D	A	B	C
500	$8.31\cdot10^{-11}$	$1.147\cdot10^{-8}$	$6.24\cdot10^{8}$	$2.394\cdot10^{4}$	$1.357\cdot10^{7}$	$8.20\cdot10^{8}$
600	$5.61\cdot10^{-7}$	$1.692\cdot10^{-5}$	$6.09\cdot10^{5}$	$7.96\cdot10^{2}$	$6.05\cdot10^{4}$	$8.12\cdot10^{5}$
700	$3.045\cdot10^{-4}$	$3.103\cdot10^{-3}$	$4.304\cdot10^{3}$	$7.00\cdot10$	$1.267\cdot10^{3}$	$5.74\cdot10^{3}$
800	0.03426	0.1547	104.9	11.30	69.7	139.9
900	1.350	3.234	5.84	2.736	7.311	7.79
1000	25.50	36.81	0.579	0.880	1.203	0.772
1100	282.3	269.3	0.0874	0.3477	0.2750	0.1166

The step-by-step procedure starts with an arbitrary value for y_4, e.g., 0.1 for (9.55) or 0.0001 for (9.57); in general, a guess of the order of magnitude is sufficient. The starting value is introduced on the right hand side of the equation and furnishes a second estimate of y_4, which in turn is introduced into the equation used. The procedure is repeated until the results of several subsequent steps are equal within acceptable limits of error.

But the procedure is not necessarily convergent. A negative result for y_4 or y_4^4 clearly indicates divergence. If this happens one can try to enforce convergence by dampening, i.e., one chooses a dampening factor F between 0 and 1, and estimates the mth approximation $y^{(m)}$ from the preceding ones $y^{(m-1)}$ and $y^{(m-2)}$ as

$$y^{(m)} = Fy^{(m-1)} + (1-F)y^{(m-2)}\ . \tag{9.58}$$

In the present case one can also try to reduce the last estimate of y_4 in order to achieve convergence. The success of such remedies is never assured. In programming such devices one should include print-outs recording intermediate steps since there is a danger of arriving at a pseudo-convergence in a wrong place.

Approximation by means of (9.57) led to results for 500 K and 600 K; use of (9.55) gave results for 1000 K and 1100 K (see Table 9.7).

Table 9.7

Boundary of Carbon Deposition

T(K)	w	w/(1-w)	n_1	n_2	n_3	n_4	n_5	$n_3/(1-w)$
500	0.212	0.269	0.783	0.212	0.0166	$2.64 \cdot 10^{-6}$	0.00428	0.021
600	0.358	0.558	0.623	0.316	0.0853	$1.91 \cdot 10^{-4}$	0.0211	0.133
700	0.496	0.984	0.440	0.372	0.251	$3.97 \cdot 10^{-3}$	0.0598	0.500
800	0.585	1.41	0.271	0.341	0.526	$3.58 \cdot 10^{-2}$	0.104	1.28
900	0.588	1.42	0.153	0.230	0.876	0.159	0.100	2.12
1000	0.528	1.11	0.0809	0.0960	1.215	0.352	0.0401	2.57
1100	0.500	1.00	0.0390	0.0300	1.392	0.453	0.0093	2.78

As soon as y_4 is found, one computes the other mole fractions by means of (9.50) to (9.53). Dividing (9.45) by n one obtains

$$n = 1/(1 - 2y_4 - 2y_5) \tag{9.59}$$

and w follows from (9.42) or (9.43) and (9.46).

When both approximation procedures fail, as here between 600 and 1000 K, we could solve (9.55) by some other method. For instance, one could calculate the right hand side for a set of arbitrary values of y_4 and plot the difference between the results and the assumed y_4's against y_4. The zero value of the difference furnishes the desired value of y_4. But the following discussion will show a different method, which sometimes may be useful.

9.43 Cyclic Sequence. The algebraic elimination of all but one unknown as performed in the preceding example may be tedious or even impossible. It can be avoided in a cyclic sequence of equations equivalent to the step-by-step procedure to be used for(9.55). As an example we shall devise a cycle to be used in a method which furnishes more extensive information of the equilibria involved. We shall at first disregard the deposition condition (9.47) and compute the mole fractions of the equilibrium mixture as functions of the mole fraction w of water in the feed. Plotting the quantity y_5/y_4^2 as a function of w we shall immediately see the border point prescribed by K_D between absence and deposition of carbon.

We choose the following equations, namely, from (9.49)

$$n_3 = (K_2 n_1 n_2^2 n^2/n_5)^{0.25} , \tag{9.60}$$

from (9.44), (9.46), (9.48), (9.49)

$$n_5 = n_3/(4 + 3K_1 n_3/K_2 n_2) \tag{9.61}$$

$$n_4 = K_1 n_3 n_5/K_2 n_2 , \tag{9.62}$$

from (9.42) and (9.43)

$$n_1 = 1 - w - n_4 - n_5 \tag{9.63}$$

$$n_2 = w - n_4 - 2n_5 \tag{9.64}$$

$$n = 1 + 2n_4 + 2n_5 . \tag{9.65}$$

For the start one has to choose values for the quantities required by (9.60), for instance,

$$n_1 = 0.95 - w; \quad n_2 = w - 0.1; \quad n_5 = 0.05; \quad n = 1.1. \tag{9.66}$$

It does not matter which values are chosen, except that none must be zero or negative (the present choice restricts w to $0.1 < w < 0.95$). With these initial values the sequence (9.60) to (9.65) is repeated until the difference between several subsequent results is tolerably small.

Equation (9.60) is selected as the first step of the cycle because n_3, appearing as a fourth root, is less sensitive to deviations of the other unknowns than any of the other quantities.

A negative value of either n_1 or n_2 appearing in (9.63) or (9.64) in the $(m+1)^{st}$ cycle can be taken as a signal of divergence. One may try to force convergence in various ways. For instance, one reduces in this case n_4, n_5 in view of (9.63) and (9.64)

$$n_4^{(m)} = 0.7\, n_4^{(m-1)}; \quad n_5^{(m)} = 0.7\, n_5^{(m-1)} . \tag{9.67}$$

One has to reduce both because either could be the cause of a negative value of n_1 and n_2. At the same time one may dampen the procedure in the manner indicated by (9.58).

This procedure furnished the values of $y_5/y_4^2 = n_5 n/n_4^2$ as a function of w for 500 to 900 K as shown in Fig. 9.4. Carbon may be deposited if the water content of the feed is below the intersection of K_D and y_5/y_4^2. Table 9.7 contains the results thus obtained for 500 to 900 K. The values for 500 and 600 K agree with the results of (9.57).

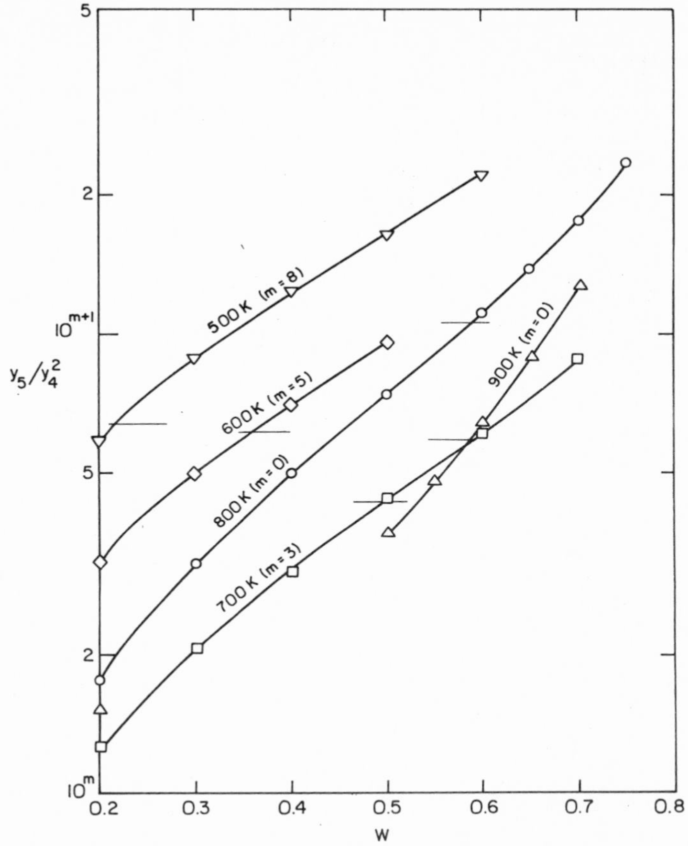

Fig. 9.4. Determination of the point of carbon deposition (the ratio y_5/y_4^2 is plotted on a logarithmic scale. The equilibrium constants K_D are indicated by short horizontal lines)

The last column of Table 9.7 shows the potential yield of hydrogen in moles per mole feed methane. The calculated ratio steam/methane (Table 9.7 and Fig. 9.5) may be questioned for the highest temperatures because of possibly disturbing side reactions.

9.44 Free Energy Minimum. The progress of automatic computing has kindled interest in the most direct method of determining a chemical equilibrium: the search for the minimum of the Gibbs free energy. One has to represent the total free energy algebraically as a function of the numbers of moles (at constant temperature and pressure) and to find its minimum with the stoichiometric relations as constraints.

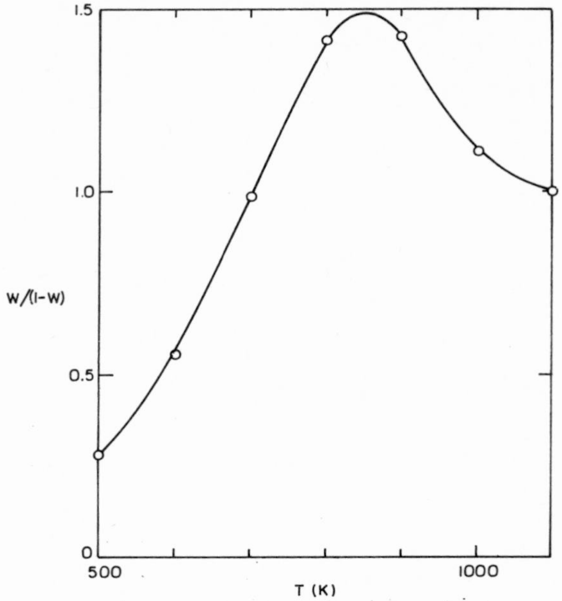

Fig. 9.5. Steam-methane ratio for the deposition point

This method has some attractive features. Computer methods for finding a minimum of a function of several variables are highly developed. A far-reaching mechanization with a very wide scope appears to be possible in connection with the use of standardized thermodynamic tables.

Any method depends of course on the availability and quality of the experimental data used, including activity coefficients and fugacity coefficients. But the minimum search depends also on the quality of the algebraic representation of the data.

9.5 EXPLOSION LIMITS

Any unexpected spontaneous explosion of large amounts of stored materials leads to various, obviously important questions. The first is of course the cause and mechanism of such disastrous explosions. The second problem is the recognition of any potential explosion danger for a given substance. Finally for mixtures one wants to find out the border between safe and unsafe concentration ranges.

Disastrous explosions of ammonium nitrate have been suspected to be caused by oxides of nitrogen or by nitrite, produced by small amounts of reducing substances such as paper or burlap. An explosion of a mixture of acetic acid and hydrogen peroxide may be explained by the intermediate formation of monoperoxyacetic acid, which is known to be more reactive than the fresh mixture of the components. Whatever the cause and the kinetics may be, there remains the question of the safety of a specified substance. To some extent it can be answered by thermodynamic computation.

If we list all conceivable reactions of a given substance and find that the corresponding increases ΔG of the Gibbs free energy are positive we may safely conclude that there is no explosion danger. The converse argument is not to be taken strictly. A serious danger does not necessarily exist if ΔG is negative but only if $-\Delta G$ is large and if also the change ΔH of the heat content has a large negative value. This is of course a somewhat vague criterion. Between reliable safety and definite danger there is a zone of uncertainty, but the thermodynamic calculation furnishes at least a useful guide for experimental investigations.

As an example we may examine the system acetic acid-hydrogen peroxide-water. Organic acids are used as oxidation catalysts and as solvents. The explosion limits of hydrogen peroxide mixtures have been investigated by Monger, Sello and Lehwalder [95]. Values of ΔG and ΔH have been computed by Mrs. Ngo Sohn [96]. One has to take into account three reactions

$$\text{(A)}\quad \underset{(1)}{CH_3COOH(\ell)} + \underset{(2)}{4H_2O_2(\ell)} = \underset{(3)}{2CO_2(g)} + \underset{(4)}{6H_2O(g)} \tag{9.68}$$

$$\text{(B)}\quad \underset{(1)}{CH_3COOH(\ell)} + \underset{(2)}{2H_2O_2(\ell)} = \underset{(5)}{2CO(g)} + \underset{(4)}{4H_2O(g)} \tag{9.69}$$

$$\text{(C)}\quad \underset{(2)}{H_2O_2(\ell)} = \underset{(4)}{H_2O(g)} + \underset{(6)}{0.5\ O_2} \tag{9.70}$$

and, in addition, the vaporization of water and acetic acid

$$\text{(D)}\quad H_2O(\ell) = H_2O(g) \tag{9.71}$$

$$\text{(E)}\quad CH_3COOH(\ell) = CH_3COOH(g)\ . \tag{9.72}$$

We assume completion of these "basic" reactions and complete gasification of the liquid mixture. It is further assumed that the combined reaction for any point of the stoichiometric diagram of Fig. 9.6 is the sum of three basic reactions represented by the nearest three of the points A, B, ... E. The participation factors can then be computed stoichiometrically. For a point in triangle ACD, for instance, with the coordinates $x_1(CH_3COOH)$, $x_2(H_2O_2)$ and $x_4(H_2O)$ the participation factors a, c, d for the reactions A, C, D follow from

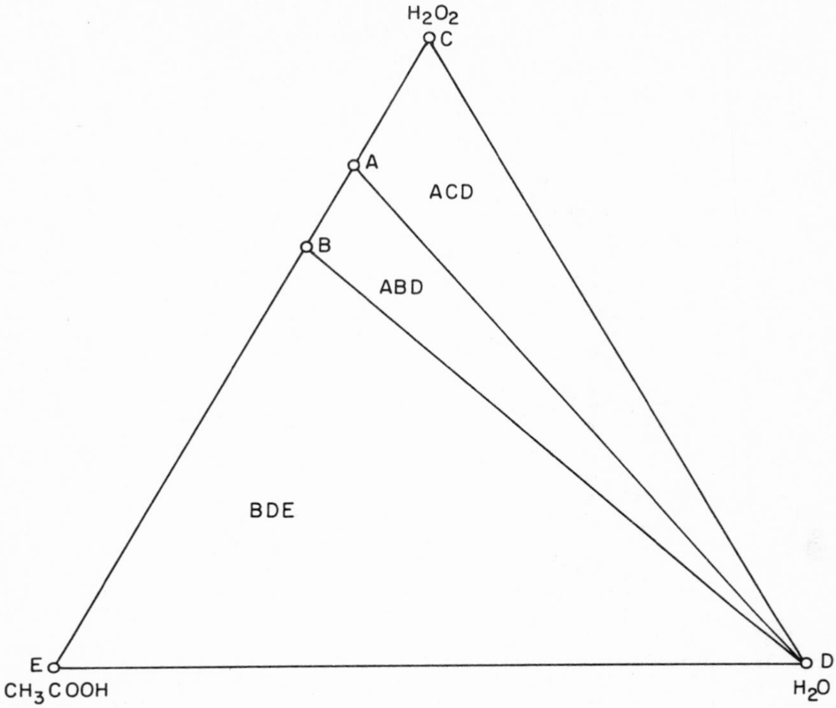

Fig. 9.6. Stoichiometric diagram CH_3COOH-H_2O_2-H_2O (mole fractions)

$$\lambda x_1 = a; \qquad \lambda x_2 = 4a + c; \qquad \lambda x_4 = d \tag{9.73}$$

so that

$$a = \lambda x_1; \qquad c = \lambda(x_2 - 4x_1); \qquad d = \lambda x_4 \tag{9.74}$$

or, referred to the reaction of 1 mole acetic acid,

$$a = 1; \qquad c = x_2/x_1 - 4; \qquad d = x_4/x_1 \ . \tag{9.75}$$

One obtains similar relations for the other fields. The sum of the ΔG and the ΔH values for the basic reactions, weighted by the participation factors a, b, ... , furnish the total values ΔG and ΔH referred to one mole of the acid. The contributions of the free energies of mixing in the two phases may be assumed to be the values of a perfect solution. Actually no serious error is committed by leaving them out since they are small compared with the chemical contributions.

The values of ΔG and ΔH in kcal/kg of the mixture are represented in Figs. 9.7 and 9.8. They have been computed by Mrs. Ngo Sohn [96].

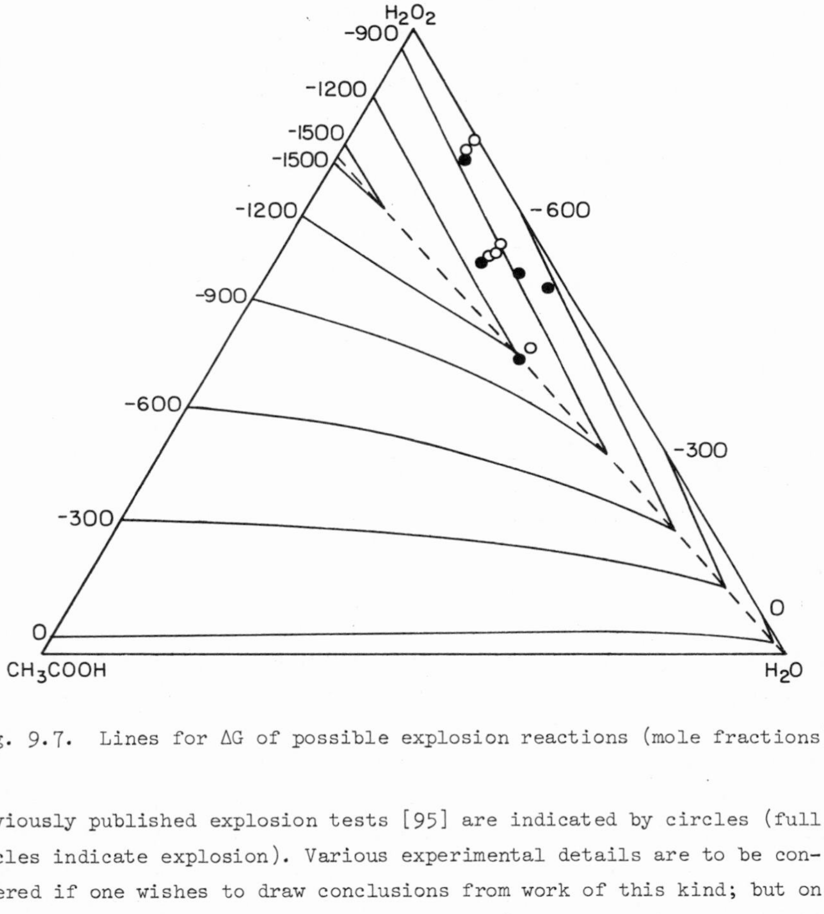

Fig. 9.7. Lines for ΔG of possible explosion reactions (mole fractions)

Previously published explosion tests [95] are indicated by circles (full circles indicate explosion). Various experimental details are to be considered if one wishes to draw conclusions from work of this kind; but one will conclude that the area enclosed by the -600 line for ΔG or the -300 line for ΔH is subject to serious danger of explosion.

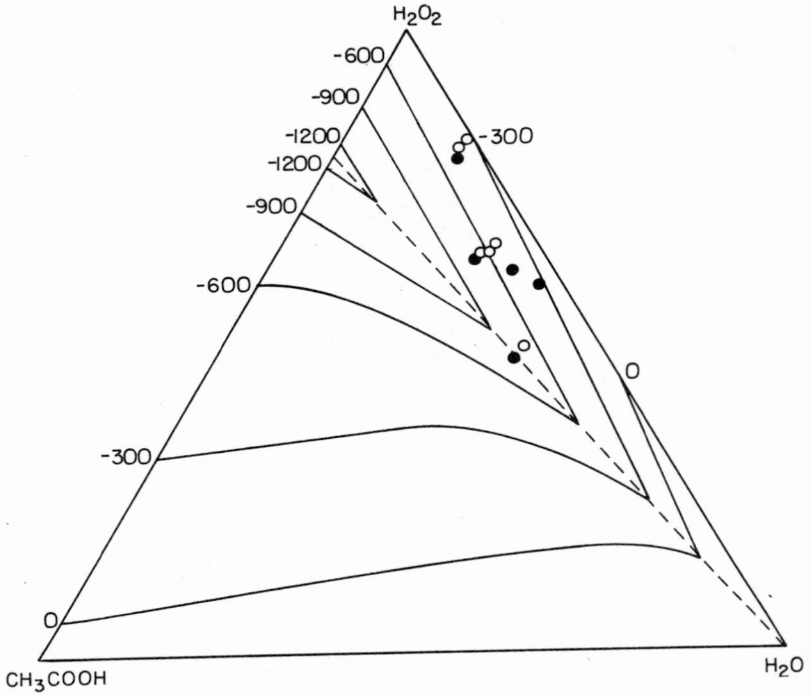

Fig. 9.8. Lines for ΔH of possible explosion reactions (mole fractions)

SUMMARY OF THE NINTH CHAPTER

In the previous chapters important tools, such as standard free energies, fugacity coefficients and activity coefficients, have been developed for the thermodynamic analysis of chemical reactions. The discussion of the equilibrium constant in this chapter concludes this development.

The application of thermodynamic methods is shown in the detailed discussion of four examples. They cover reactions of great practical importance. But the emphasis here rests on the variety of methods to be used in the investigation of technical problems. Technological knowledge should lead to flexibility and adaptability.

References

(91) C. E. Wicks and F. E. Block, "Thermodynamic Properties of 65 Elements," U. S. Bureau of Mines, Bulletin 605. Washington, 1963.

(92) A. T. Larson and R. L. Dodge, J. Am. Chem. Soc., 45 (1923) 2918.
A. T. Larson, J. Am. Chem. Soc., 46 (1924) 367.

(93) T. J. Browder (The Ralph M. Parsons Company), Lecture at the 63rd Annual Meeting, Amer. Inst. Chem. Eng., 1970.

(94) B. F. Dodge, Trans. Am. Inst. Chem. Eng., 34 (1938) 540. O. A. Hougen, K. M. Watson and R. A. Ragatz, "Chemical Process Principles" Pt. II, 2nd ed., John Wiley & Sons, New York, 1959, p. 1046.

(95) J. M. Monger, H. Sello and D. C. Lehwalder, J. Chem. Eng. Data, 6 (1961) 23.

(96) V. B. T. Ngo, M. S. Thesis, Lawrence Radiation Laboratory Report UCRL-18510, Berkeley, California, 1968.

CHAPTER 10: THE THERMODYNAMIC ROOTS OF VARIOUS PHENOMENA

In the preceding chapter our attention was concentrated on "chemical" thermodynamics; in this limited field we needed only a few kinds of generalized coordinates, mainly the volume and the numbers of moles. But in the first chapter we had recognized thermodynamics as the general doctrine of equilibrium and concluded that all physical sciences have some roots in thermodynamics.

When we wish to introduce a new field, we have to go back to the basic concepts of Chapter 1. We have to find a pertinent general coordinate, independent of the other coordinates previously introduced, and its conjugate force. The rest is a matter of applying the general relations arrived at in the second chapter.

10.1 SURFACE PHENOMENA

Just as in Chapter 1 we consider a very simple example (Fig. 10.1)

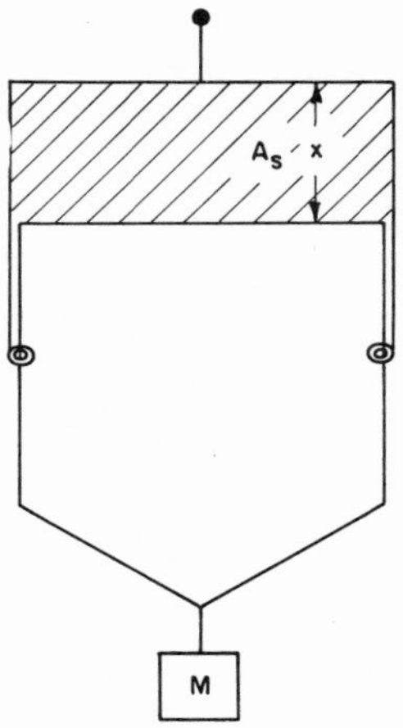

Fig. 10.1. Surface tension (M mass, L length of the edge, A_s surface area, x width)

in order to introduce a variable which satisfies the requirements of a generalized coordinate. It is the <u>interface area</u> between two phases. We call it <u>surface area</u> or simply surface if one of the phases is gaseous (though this distinction is not always strictly maintained). As a rule, the interface area can certainly be chosen as an independent variable: the surface of a given amount of mercury increases if we pour it from a beaker into a large tray. Clearly we can keep the surface area constant and we can change it according to our pleasure without changing any of the other coordinates. Therefore, the interface area is a suitable generalized coordinate characteristic for surface phenomena.

It is an <u>extensive</u> quantity (Section 1.31). Mercury in two identical trays presents twice the surface area.

What we call surface is really a transition region; usually it involves only a few molecular layers. In general, a thorough examination of the structure of an interface is required in molecular theory; but there is no need to discuss it here.

Without giving the transition region any special thought, we can always divide up the molecules of a pure substance between two phases, say liquid and vapor, so that at the same time the total volume is divided up in the volumes of two phases of the proper densities. But we cannot do the same for two or more components since the ratios of the concentrations in the bulk phases and in the transition region may be(and in general are) different. For this reason one adopts the convention that the amount of the first component (solvent) is completely divided up between the bulk phases so that nothing remains for the interface; then definite amounts of the other components remain which have to be assigned to the surface. Thus we have surface concentrations Γ_i expressed, for instance, in moles/cm^2; they indicate the enrichment of the surface in the components concerned. For the solvent we have $\Gamma_1 = 0$. The surface may also be impoverished of some component so that $\Gamma_i < 0$.

10.11 <u>General Relations</u>. The generalized force conjugate to the surface area can be recognized by inspection of the interaction of a surface with a weight (Fig. 10.1). A soap film is spanned between horizontal bars of a fixed wire frame and a mobile one. The film is stretched by the weighted mobile frame until equilibrium is reached between the force exerted by the weight and the counterforce exerted by the surface film along the edge of the mobile frame and perpendicularly to it. The counterforce acting on the unit length of the frame edge is called surface tension σ (or interfacial tension). The equilibrium condition is expressed by the gravitational force Mg (mass times gravitational constant) of the weight and the length L of the edge (twice for the front and back surfaces) as

$$2L\sigma = Mg. \tag{10.1}$$

The surface tension is therefore a generalized force of the dimension force/length; it has been traditionally expressed in dyne/cm (in the International System of 1960 replaced by newton/meter, 1 Nm^{-1} being equal to 10 dyne/cm).

The reversible work w_s of extending a surface by the area dA_s in the direction x perpendicular to the edge is therefore (Eq.(1.13))

$$w_s = \sigma dA_s. \tag{10.2}$$

This is now a term in the work sum introduced in the general equation (2.32).

We include in the discussion several phases and transfer of a component i into or from a phase j; the amount transferred will be called dn_i^j. The generalized force conjugate to n_i^j is the chemical potential or the partial molal free energy, which we write now

$$\mu_i = \bar{G}_i^{(1)} = \bar{G}_i^{(2)} = \ldots \tag{10.3}$$

implying that (in equilibrium) the value of the chemical potential is equal in all phases. Thus we obtain finally from (2.32), (2.33) and (2.38)

$$dE = TdS - PdV + \sigma dA_s + \sum_{i,j} \mu_i \, dn_i^j \tag{10.4}$$

$$dA = -SdT - PdV + \sigma dA_s + \sum \mu_i \, dn_i^j \tag{10.5}$$

$$dG = -SdT + VdP + \sigma dA_s + \sum \mu_i \, dn_i^j \,. \tag{10.6}$$

The second summation symbol j in Eq.(10.4) indicates that one has to sum not only over all z components i in every phase but also over all phases j present. The notation is simplified in the second and third equations.

A generalization of the Eq.(1.41) of Gibbs is obtained in the following manner. We assemble a large number of identical objects, each characterized by the values dE, T, dS, P and so on of Eq.(10.4). This is a purely fictional process; no specific interaction or reaction takes place since the objects have the same temperature, pressure and so on. The small quantities dE, dS, dV, dA_s, dn_i^j are all extensive and therefore add up to large quantities E, S, V and so on; the quantities T, P and so on are intensive and remain unchanged. The result of this integration is

$$E = TS - PV + \sigma A_s + \sum_{i,j} \mu_i n_i^j \,. \tag{10.7}$$

Now we differentiate without any restriction and have

$$dE = TdS + SdT - PdV - VdP + \sigma dA_s + A_s d\sigma + \sum_{i,j}(\mu_i dn_i^j + n_i^j d\mu_i^j). \qquad (10.8)$$

Subtracting (10.4) from (10.8) we obtain the extended equation of Gibbs

$$SdT - VdP + A_s d\sigma + \sum_{i,j} n_i^j d\mu_i^j = 0. \qquad (10.9)$$

This relation again is not the consequence of special thermodynamic interrelations but simply of the fact that the energy in Eq.(10.4) has been represented as a homogeneous function of first degree of the set of independent variables. It is not necessary that the variables are only the numbers of moles n_i^j. Relations of this kind hold not only for E, A, G but for any extensive property.

The wire frame of Fig. 10.1 furnishes a good illustration of area and surface tension as generalized coordinate and force, but for precise measurement different methods are used such as the capillary lift or the volume of the falling drop.

In the following only a few aspects and applications in the wide field of surface chemistry will be briefly discussed.

10.12 A Single Substance. Heat effects connected with the change of surface areas are often small but they are not always negligible.

For an interface between phases of a single component, Eqs.(10.7) and (10.9) are simplified to

$$E_s = TS_s + \sigma A_s \qquad (10.10)$$

$$S_s dT + A_s d\sigma = 0 \qquad (10.11)$$

because we do not ascribe a volume to an interface, and because we ascribe numbers of moles only when more than one component is present.

The heat content or energy content of the unit area is therefore, according to (10.10)

$$H_s/A_s = E_s/A_s = \sigma + TS_s/A_s. \qquad (10.12)$$

The interface entropy per unit area follows from Eq.(10.11) as

$$S_s/A_s = -d\sigma/dT \qquad (10.13)$$

so that

$$H_s/A_s = \sigma - T\cdot d\sigma/dT\ . \qquad (10.14)$$

The analogy with other heat content relations is obvious. Indeed, these two equations are special cases of Eqs.(3.4) and (3.8).

At the critical point the surface tension between liquid and vapor must disappear because the two phases become identical. The empirical relation [101]

$$\sigma = \sigma^\circ(1 - T/T_c)^{\frac{11}{9}} \tag{10.15}$$

complies with this requirement.

In general, the states of two phases in equilibrium are independent of the extent and properties of the interface between them. But this is not true any more if one phase is present as small droplets (say, with a radius of a fraction of a millimeter) or small bubbles. The surface tension, though, is practically the same as in flat interfaces as long as the radius is large compared with molecular dimensions. However, the interface area A_s is not an independent variable any more but linked with the state variables of the disperse phase. It can be used to describe the size of a droplet just as well as its volume V, its radius r, or the number of moles n in the droplet. Thus a change of size can be represented equally well by dr or dn or

$$dV = \bar{V}dn = 4r^2\pi dr \tag{10.16}$$

or

$$dA_s = 8\pi r dr = 2dV/r = 2\bar{V}dn/r. \tag{10.17}$$

The work that must be done in introducing dn moles from a bulk phase (pressure P°) into the droplet (pressure P) can be expressed equally well by the work required for the volume change, namely,

$$(P - P^\circ)dV = (P - P^\circ)\bar{V}dn \tag{10.18}$$

or by the work required for the change of the surface area

$$\sigma dA_s = (2\sigma\bar{V}/r)dn\ . \tag{10.19}$$

The comparison of (10.18) and (10.19) shows that the droplet is not in equilibrium with the bulk phase but is under a higher pressure, namely,

$$P = P^\circ + 2\sigma/r\ . \tag{10.20}$$

The pressure difference usually is small. If σ is expressed in dyne/cm and r in cm, it amounts to

$$2\sigma/r \text{ dyne/cm}^2 = 1.97 \times 10^{-6}\ \sigma/r \text{ atm} \tag{10.21}$$

or for water at 25°C with σ = 71.97 dyne/cm

$$P = P^\circ + 1.42 \times 10^{-4}/r \text{ atm.} \tag{10.22}$$

The variation of the chemical potential with pressure is given with sufficient accuracy by

$$\mu = \mu^\circ + \frac{\partial\mu}{\partial P}(P - P^\circ) = \mu^\circ + \bar{V}(P - P^\circ) \tag{10.23}$$

or for water at 25°C

$$\mu - \mu^\circ = 18.02 \times 1.42 \times 10^{-4}/r \text{ cm}^3\cdot\text{atm} \tag{10.23a}$$

so that the change in vapor pressure at room temperature can be expressed by

$$\begin{aligned} \ell n(p/p^\circ) &= (\mu - \mu^\circ)/RT = 18.02 \times 1.42 \times 10^{-4}/(82.06\cdot 298.15)(1/r) \\ &= 1.05 \times 10^{-7}/r \end{aligned} \tag{10.24}$$

if the deviation of the vapor from a perfect gas may be neglected.

This relation of Kelvin shows the increase of the vapor pressure with decreasing radius of the droplet. The effect is very small for water droplets of 1 mm radius. But for tiny dew droplets, say $r = 1$ micron $= 10^{-4}$ cm, it amounts to 0.1% and it becomes large for smaller droplets long before the size of a single molecule is reached. This effect explains the instability of small droplets in the presence of larger ones. On the other hand, very small droplets of equal size (fog) often can exist a long time before final condensation.

10.13 Adsorption of a Pure Gas. Adsorption on a solid surface has been the basis of important methods of separation and purification. Several theoretical and empirical equations have been proposed for the dependence of the adsorbed amount on the pressure of the gas. In the following discussion that gas is supposed to be perfect. As a measure of the adsorbed substance one chooses usually the amount q of adsorbate in moles or in grams per gram of solid adsorbent since it can be determined directly; the surface area can be measured, but only by indirect methods.

For low pressures Langmuir's relation [102] is well founded. Here it is assumed that the surface is saturated by the amount q_s of a monomolecular layer but that at pressures below the vapor pressure of the adsorbate only the fraction q/q_s is covered. There exists a dynamic equilibrium between the molecules evading from the surface and the molecules hitting a bare place of the surface and sticking there. The number of the evading molecules $\lambda' q$ is some constant fraction of the number of adsorbed molecules. The number of incomers is proportional to the gas pressure and to the uncovered area $(1 - q/q_s)$ so that in equilibrium

$$\lambda' q = \lambda'' P(1 - q/q_s) \ . \tag{10.25}$$

Replacing λ''/λ' by the constant

$$B = \lambda''/(\lambda' q_s) \tag{10.26}$$

we obtain Langmuir's adsorption isotherm

$$q = q_s \frac{BP}{1 + BP} \tag{10.27}$$

in which B is a measure of adsorption strength of the surface for the gaseous substance.

This relation is a limiting law for low pressures. For high pressures an empirical relation

$$q = q_s P^{1-g} \qquad (0 < g < 1) \tag{10.28}$$

proposed by Freundlich has often been found useful. A combination [103] of the two relations

$$q = q_s BP/(1 + BP^g) \tag{10.29}$$

has the advantage to approximate Eq.(10.27) for low pressures and (10.28) for high pressures. The adsorption of n-pentane on zeolite (Molecular Sieves Type 5A) is shown in Fig. 10.2. In this logarithmic diagram Freundlich's isotherm (10.28) is represented by any straight line. Langmuir's isotherm for 198°C is indicated by a broken line. The full lines represent Eq.(10.29).

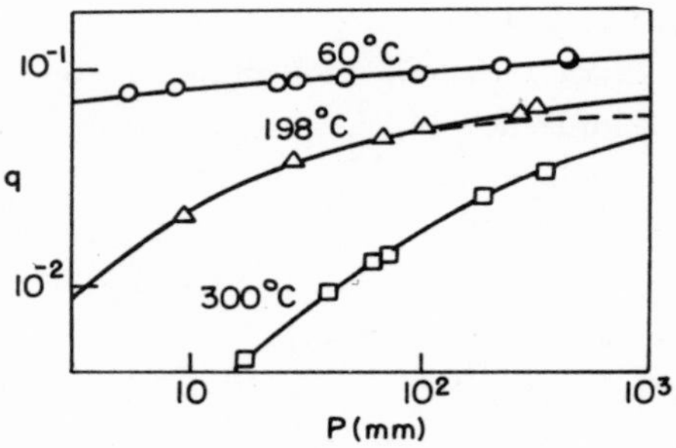

Fig. 10.2. Adsorption isotherms (n-Pentane on Molecular Sieves, Type 5A)

A relation for multilayered adsorption, valid without restriction of pressure, has been derived by Brunauer, Emmett and Teller [104] (BET). It combines the good representation of observed data with a sound theoretical basis; like Eq.(10.29), it introduces three empirical coefficients, but it is slightly more complicated.

The heat of adsorption is sometimes quite large; it can be measured calorimetrically. It follows also from the temperature dependence of the pressure at constant surface coverage. If the gas is perfect, one derives for the molal heat content of the adsorbate

$$\bar{H}_s = \bar{H}_{ads} - \bar{H}_{gas} = RT^2\left(\frac{d\ell nP}{dT}\right)_q \tag{10.30}$$

in precisely the same way in which Clapeyron's equation is obtained. For Langmuir's relation (10.27) one finds immediately

$$[d(BP)/dT]_q = 0; \qquad (d\ell nP/dT)_q = -d\ell nB/dT. \tag{10.31}$$

The corresponding equations for other adsorption isotherms are more complicated.

10.14 Surface Concentration. In the presence of a mixture, an interface may be enriched in one of the components. The enrichment Γ_i can be measured (moles of component i/surface area) and is a thermodynamically significant quantity. It is discussed in the following for the simplest case, a solution in a nonvolatile solvent.

We specialize Eq.(10.6) for this case and for constant temperature, pressure and total number of moles $[n_1(\ell) + n_1(q)]$ of the solvent (component 1) in the two phases. Thus we have for a change of the total Gibbs free energy

$$dG = \sigma dA_s + \mu_2 dn_2 \tag{10.32}$$

$$\left(\frac{\partial G}{\partial A_s}\right)_{T,P,n_1,n_2} = \sigma\ ; \qquad \left(\frac{\partial G}{\partial n_2}\right)_{T,P,A_s,n_1} = \mu_2\ .$$

We shall not explicitly indicate the constancy of T, P, n_1 in the following. Cross differentiation in the usual manner furnishes

$$\frac{\partial^2 G}{\partial A_s \partial n_2} = \left(\frac{\partial \sigma}{\partial n_2}\right)_{A_s} = \left(\frac{\partial \mu_2}{\partial A_s}\right)_{n_2} \ . \tag{10.33}$$

The last term means the change of the chemical potential of the solute when its total number of moles is kept constant while the surface area is changed. We exclude small bubbles so that the potential does not depend explicitly on the surface area; but the increase dA_s of the area at constant total n_2 entails a reduction of the amount of solute in the liquid, which can be expressed by means of the enrichment or surface concentration Γ_2 of the solute as

$$dn_2(\ell) = -\Gamma_2 dA_s\ . \tag{10.34}$$

The surface concentration is therefore, in view of Eq.(10.33),

$$\Gamma_2 = -\left(\frac{dn_2(\ell)}{dA_s}\right)_{n_2} = -\frac{dn_2(\ell)}{d\mu_2}\cdot\left(\frac{\partial \mu_2}{\partial A_s}\right)_{n_2} = -\left(\frac{\partial \sigma}{\partial n_2}\right)_{A_s}\cdot\frac{\partial n_2}{\partial \mu_2} \tag{10.35}$$

We omit the phase indication in the last term because μ_2 depends only on $n_2(\ell)$. Replacing the potential by the activity coefficient

$$\mu_2 = RT\ell n(x_2\gamma_2) \tag{10.36}$$

we obtain

$$\Gamma_2 = -\frac{1}{RT}\cdot\frac{d\sigma}{d\ell n(x_2\gamma_2)}\ . \tag{10.37}$$

The activity $x_2\gamma_2$ is a monotonic function of x_2. Therefore the solute is enriched in the surface if $d\sigma/dx_2$ is negative. If the surface tension increases with the concentration, the surface is impoverished in the solute.

10.15 Selective Adsorption. Purification by means of solid adsorbents has been a long established method in the laboratory and in the plant. In recent years adsorption methods have been rapidly developed by the introduction of new adsorbents [105] (e.g., R. M. Barrer's "Molecular Sieves") and new operation techniques.

The separation factor (7.2), here called relative adsorbability, is again a measure of the efficiency of separation.

The adsorption equations (10.27) and (10.29) are easily extended to mixtures [106]. Instead of (10.29) one has for a perfect gas mixture (y_i = mole fraction)

$$q_i = A_i y_i P/[1 + \sum_j B_j (y_j P)^{g_j}] \tag{10.38}$$

and Langmuir's isotherm is given by Eq.(10.38) with $g_j = 1$. An example is shown in Fig. 10.3.

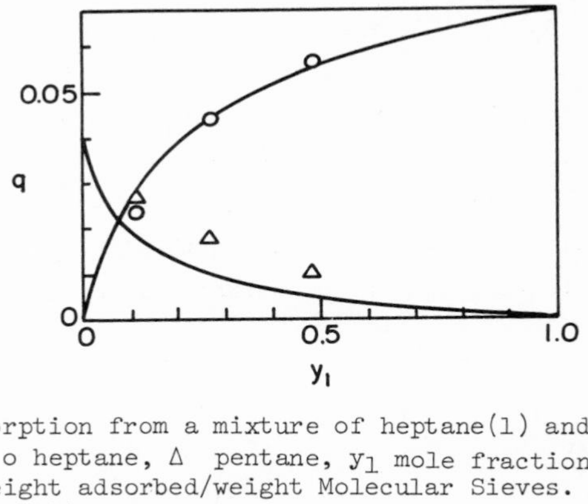

Fig. 10.3. Adsorption from a mixture of heptane(1) and pentane(2). (300°C, 550 mm, o heptane, Δ pentane, y_1 mole fraction of heptane in the gas, q weight adsorbed/weight Molecular Sieves. The curves represent Eq.(10.38))

With the aid of Eq.(10.38) the relative adsorbability can be approximated by

$$\alpha_j^i = \frac{q_i}{q_j} \cdot \frac{y_j}{y_i} = \frac{A_i}{A_j} \tag{10.39}$$

if a common value can be used for all exponents g_j.

Relations such as (10.38) and (10.39) can easily be transformed into an equation for the adsorption from perfect liquid solutions with the aid of the vapor pressures p_i^o and the mole fractions x_i in the liquid; one has only to replace

$$y_i P = x_i p_i^o \; . \tag{10.40}$$

The general principles for the application of adsorption are similar to those for extraction. Chromatographic columns, early introduced by Tswett and developed by Zechmeister, were the prototype of multistage adsorption. An extensive development of adsorption techniques started from Martin and Synge's investigation [107] of the efficiency of adsorption columns.

Adsorption is often favored over extraction by very high values of the relative adsorbability. Moreover, there are different adsorbents specifically useful for different separations. Charcoal adsorbs nonpolar substances, separating primarily according to molecule size. Silica preferentially selects according to polarity. Zeolites (Molecular Sieves) are also selective adsorbents for polar molecules but their primary value is the extremely sharp selectivity according to size and shape of the adsorbate molecule.

Even if the relative adsorbability is not far from unity, a high efficiency can be attained in an adsorption column since it can be designed to be equivalent to a very large number of stages (adsorptions and desorptions).

The inherent disadvantage of adsorption columns is the low capacity. Only in rare cases does an adsorbent hold as much adsorbate as 10% or 20% of its own weight. The high selectivity for an adsorbate necessarily causes problems in the desorption and in the regeneration of the adsorbent.

Though large-scale adsorption methods have been developed, analytical and biochemical problems have been the most favorable field, in which chromatographic applications have astonishing success. Ultra-micro methods as required in clinical chemistry, biological investigations or space chemistry are not hampered by the capacity limitations of adsorption columns.

10.2 CENTRIFUGAL SEPARATION

The great power of centrifugal processes rests on the fact that we can produce forces that are similar to gravitational forces but very much greater. The force exerted by the earth on the mass M is Mg with the standard gravity being $g = 980.665\ \text{cm}\cdot\text{sec}^{-2}$. The centrifugal force at a distance r from the axis of a centrifuge spinning with an angular velocity ω is

$$f = -M\omega^2 r. \tag{10.41}$$

For an ultracentrifuge operating at 30 000 r.p.m. the angular velocity is approximately $2\pi\cdot 30\,000/60 = 3100\ \text{sec}^{-1}$ and at a radial distance of 1 cm the centrifugal force is equivalent to about 10 000 Mg.

For a long time centrifuges have been used for industrial phase separations (solid-liquid, solid-gas, liquid-liquid). In such cases the high cost of the equipment may be balanced by a very high through-put as well as by efficiency of separation. But the application to the separation of molecules on the basis of different molal weights required the design of small centrifuges spinning at extremely high speeds. An advantage in increasing ω rather than r is suggested by Eq.(10.41); small size, moreover, is necessary to attain the very high strength of the rotator required to withstand a high centrifugal force.

The coordinate conjugate to the centrifugal force is the distance r from the axis so that we can take centrifugation into account by introducing the work

$$-M\omega^2 r\ dr \tag{10.42}$$

as one of the terms in the sum occurring in Eqs.(2.32), (2.33) and (2.38). The negative sign in Eqs. (10.41) and (10.42) expresses the fact that the object's energy decreases with increasing r. In the same way as Eq.(10.6) we obtain

$$dG = -SdT + VdP - M\omega^2 rdr + \sum \mu_i dn_i\ . \tag{10.43}$$

The indication of the phase is not necessary here since we have only a single phase (or, if we prefer, a continuous sequence of phases). In any case, the variables in Eq.(10.43) may vary with the distance r.

In the discussion of surface phenomena the coordinate A_s was an independent variable with the exception of the case of small droplets, where it was linked to the pressure. In the centrifuge, however, pressure always increases in the interval dr by the centrifugal force of the corresponding layer. If the cross section of the centrifuge is called q and its density

$$\rho = M/V, \tag{10.44}$$

the volume and the mass of the layer are given by

$$dV = q\ dr \tag{10.45}$$

$$dM = \rho\ dV\ . \tag{10.46}$$

The increase of the pressure is therefore

$$dP = \omega^2 r dM/q = \rho\omega^2 r dV/q = \rho\omega^2 r dr\ . \tag{10.47}$$

Eliminating dP from Eq.(10.43) and specializing for isothermal changes we have

$$dG = (V\rho - M)\ \omega^2 r dr + \sum \mu_i dn_i\ . \tag{10.48}$$

Cross differentiation leads to

$$\partial^2 G/\partial r \cdot \partial n_i = (\bar{V}_i\rho - \bar{M}_i)\omega^2 r = (\bar{v}_i\rho - 1)\bar{M}_i\omega^2 r$$

$$= (\partial\mu_i/\partial r)_{n_a}\ . \qquad (n_a \neq n_i) \tag{10.49}$$

Here we introduce the partial molal volume $\bar{V}_i$ and the molal weight $\bar{M}_i$, connected with the total mass M by

$$M = \sum n_i\bar{M}_i \tag{10.50}$$

and we abbreviate by introducing the partial specific volume of component i

$$\bar{v}_i = \bar{V}_i/\bar{M}_i\ . \tag{10.51}$$

In the following the composition of a solution of z components will be represented by the mole fractions x_2, x_3 ... x_z. The mole fraction x_1 of the solvent is given by the other mole fractions, and its chemical potential by the relation of Gibbs and Duhem.

The experimental procedure for equilibrium measurements consists of spinning a centrifuge at constant temperature and speed until the final composition of the solution (determined as a function of r by means of the refractivity or by other means) is reached. Then diffusion has stopped and the chemical potential μ_i of all components must have the same value everywhere in the centrifuge.

The equilibrium condition is therefore

$$\frac{d\mu_i}{dr} = \left(\frac{\partial\mu_i}{\partial r}\right)_{x_a} + \sum_a \left(\frac{\partial\mu_i}{\partial x_a}\right)_r \cdot \frac{dx_a}{dr} = 0\ . \tag{10.52}$$

We use Eqs. (10.49) and (10.51) and introduce the activity $x_i\gamma_i$ so that

$$\frac{d\mu_i}{dr} = \bar{M}_i\omega^2 r(\rho\bar{v}_i - 1) + RT\sum_a \frac{\partial \ell n(x_i\gamma_i)}{\partial x_a} \cdot \frac{dx_a}{dr} = 0. \tag{10.53}$$

The activity coefficient γ_i is defined here as elsewhere in this book for a solution at 1 atm; it is a function of the composition of the solution, i.e., of the z-1 mole fractions x_2, x_3 ... x_z (an other z-1 composition variables could be used in Eq.(10.53)). We may write therefore, after division by RT and separation of $\ell n\, x_i$ from the sum

$$\frac{M_i\omega^2 r}{RT}(\rho\bar{v}_i - 1) + \frac{d\ell n\, x_i}{dr} + \sum_a \frac{\partial \ell n\, \gamma_i}{\partial x_a}\cdot\frac{dx_a}{dr} = 0. \tag{10.53a}$$

These are z-1 equations, one for each solute i. One measures $d\ell n\, x_i/dr$ and can solve for the molal weights $\bar{M}_i$, if the activity coefficients are known or can be estimated. In this case one replaces the mole fractions by another measure of the composition not involving the molal weights, such as the weight of the component in 1000 grams of solvent. This application is important in polymers and above all in biological investigations. The pertinent literature is extensive [108].

If the molal weight is known, Eq.(10.53) furnishes information about activity coefficients. This has been successfully demonstrated by Young and his coworkers [109].

Centrifugation of a gaseous mixture has been contemplated for the separation of the hexafluorides of the uranium isotopes. We assume that the gas is perfect, so that the last term in Eq.(10.53) is zero. We introduce Eqs.(10.51) and (10.47) in the first term of (10.53) and find

$$\bar{M}_i\bar{v}_i = \bar{V}_i = RT/P; \qquad \rho\omega^2 r = dP/dr \tag{10.54}$$

$$\frac{1}{r}\cdot\frac{dP}{dr} - \frac{M_i\omega^2 r}{RT} + \frac{d\ell n\, y_i}{dr} = 0 \tag{10.55}$$

writing y_i for the mole fraction in the gas. The integration of

$$\frac{d\ell n(Py_i)}{dr} = \frac{\bar{M}_i\omega^2}{RT}r \tag{10.56}$$

furnishes, if the gas is introduced at r = 0 with the partial pressure $\dot{P}\dot{y}_i$ of i,

$$\ell n(Py_i/\dot{P}\dot{y}_i) = \frac{\bar{M}_i\omega^2}{2RT}\cdot r^2 \tag{10.57}$$

so that the separation factor is

$$\alpha_k^i = y_i\dot{y}_k/(\dot{y}_i y_k) = \exp[(\bar{M}_i-\bar{M}_k)\omega^2 r^2/(2RT)]. \tag{10.58}$$

These relations can also be used for the atmosphere if the centrifugal work $\bar{M}_i\omega^2 r^2/2$ is replaced by the gravitational work $\bar{M}_i gh$ (gravitational acceleration times height). The validity of the result is limited by the implied assumptions (equilibrium, constant temperature, gravitational acceleration independent of height, perfect gas).

10.3 ELECTROSTATICS

The basic facts of electrothermodynamics have been discussed in the first chapter: The electric charge Q has the definite qualities of a generalized coordinate and the voltage or potential difference E is its conjugate generalized force. This applies to a galvanic cell as well as to a capacitor. In the galvanic cell, however, we are usually (not necessarily) interested primarily in a chemical reaction; as a rule, therefore, we replace the electric charge by numbers of mole of the reaction partners as coordinates.

In the ideal capacitor no chemical reactions occur. We retain therefore the charge as an independent variable. Even the properties of a non-reacting and non-conducting substance, called dielectric, change under the influence of an electric field. The thermodynamics of these changes follows in a straightforward discussion starting from the generalized coordinate and force.

10.31 Molal Properties. In order to determine the electric work done upon a given amount of dielectric we shall find the work done on it in a suitable capacitor [109.1] and subtract the work required for establishing the field in the empty capacitor. The capacitor is shown in Fig. 10.4.

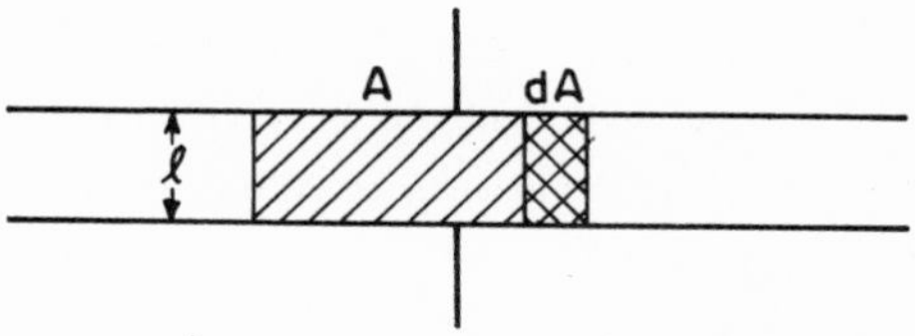

Fig. 10.4. Capacitor (ℓ fixed distance between plates, A area)

The distance ℓ between the plates is fixed. The dielectric occupies the area A between the plates of the volume

$$V = \ell A . \qquad (10.59)$$

It can expand by

$$dV = \ell dA \qquad (10.60)$$

into the empty space between the plates which had been subjected to the same field as the dielectric. We cannot admit a variable distance ℓ because the field would be changed and the dielectric would expand into field-free space.

We introduce now the conventional quantities of electromagnetic theory, namely, the electric field strength

$$e = E/\ell \tag{10.61}$$

and the charge density

$$q = Q/A = \varepsilon e . \tag{10.62}$$

The quantity ε, called the permittivity, is the property of the dielectric which controls its interaction with the electric field. It is independent of the field strength in a wide range but the following discussion does not presuppose its constancy in any way. The permittivity ε_o of the vacuum is a constant to which different systems of electromagnetic units ascribe different numerical values.

The traditional term for q, dielectric displacement, stems from Maxwell's explanation of the state of a dielectric in a field: The positive and negative constituents of the atom are pulled slightly in opposite directions and thus produce electric dipoles that counteract the impressed field. To maintain a given field strength in a dielectric, the charge density must therefore be higher than for the empty space by a factor $\varepsilon/\varepsilon_o$, which must be greater than unity. To this effect of "induced dipoles" another one has been added by Debye. In numerous molecules, for instance methanol or aminophenol, the centers of positive and negative charges do not coincide even in absence of any field; they are "permanent dipoles". In a liquid or gaseous phase they are randomly oriented but an external field produces a trend to alignment in the direction of the impressed field. The effect on the permittivity is similar to that of induced dipoles except that it is counteracted by molecular collisions and therefore decreases appreciably with increasing temperature.

The explanation by induced and permanent dipoles illustrates the fact that we may cut the link between the state of the dielectric and the tool used to produce it, i.e., the capacitor. The use of the name "dielectric displacement" for "charge density" expresses this cut.

The dimensionless ratio $\varepsilon/\varepsilon_o = D$ is called dielectric constant though it may vary at high values of the field strength. It varies appreciably with pressure and, for polar substances, with temperature.

The work done in charging a capacitor containing the volume V of a dielectric is now, with Q and E being the generalized coordinate and force,

$$dW' = EdQ = e\ell d(Aq) = ed(Vq) = ed(V\varepsilon e) \tag{10.63}$$

in view of Eqs.(10.59) to (10.62). Subtracting the work required for charging the empty capacitor we obtain for the work done on the dielectric

$$dW = ed(V\varepsilon e) - ed(V\varepsilon_o e) = \varepsilon_o e d[(D-1)Ve] . \tag{10.64}$$

Here $\varepsilon_o e$ plays the role of the generalized force f, and $(D-1)Ve$ represents the coordinate x. Thus we obtain for the free energy of Gibbs according to Eq.(2.38)

$$dG = -SdT + VdP + \varepsilon_o e d[(D-1)Ve] . \tag{10.65}$$

We introduce a new characteristic function

$$G^* = G - \varepsilon_o e^2(D-1)V . \tag{10.66}$$

Differentiation furnishes

$$dG^* = -SdT + VdP - \varepsilon_o e(D-1)Vde . \tag{10.67}$$

Cross differentiation leads to

$$\left(\frac{\partial^2 G^*}{\partial T \partial e}\right)_P = -\left(\frac{\partial S}{\partial e}\right)_{T,P} = -\varepsilon_o e\left(\frac{\partial[(D-1)V]}{\partial T}\right)_{e,P} \tag{10.68}$$

$$\left(\frac{\partial^2 G^*}{\partial P \partial e}\right)_T = \left(\frac{\partial V}{\partial e}\right)_{T,P} = -\varepsilon_o e\left(\frac{\partial[(D-1)V]}{\partial P}\right)_{e,T} . \tag{10.69}$$

These relations are valid for any fixed amount of the dielectric. We may specialize them for one mole.

A variety of other relations can be derived. As an example we differentiate Eq.(10.68) with respect to temperature and obtain according to Eq.(3.7)

$$\left(\frac{\partial^2 S}{\partial e \partial T}\right)_P = -\varepsilon_o e\left(\frac{\partial^2[(D-1)V]}{\partial T^2}\right)_{e,P} = \frac{1}{T}\left(\frac{\partial C_p}{\partial e}\right)_{T,P} . \tag{10.70}$$

10.32 <u>The Fixed Capacitor</u>. The goal of electromagnetic theory is the description of events in space. The primary goal of thermodynamics, however, is the description of material objects. The different aspect of electromagnetic theory is the reason that previous electro-thermodynamic discussions always have started from a problem which is different from the investigation of molal properties of the dielectric.

Traditional electromagnetic theory introduces a capacitor similar to Fig. 10.4, but completely immersed into the dielectric. While in the preceding discussion the effective area A was variable, it is now fixed and the volume of the capacitor

$$V_c = \ell A \tag{10.71}$$

is a constant. The substitution of Eq.(10.71) for Eq.(10.59) entails very little formal change in the subsequent derivation until Eq.(10.65) is reached. Now we have

$$dG = -SdT + VdP + \varepsilon_o eV_c d[(D-1)e] \ . \tag{10.72}$$

The meaning of V_c is clearly established. But the meaning of the other extensive quantities, G, S, V, requires special consideration. They do not refer to a certain amount of dielectric because the amount occupying the volume V_c between the capacitor plates is variable. Yet their meaning is perfectly definite. They represent the total quantities for the whole dielectric.

As such they appear in the relations of Drude and Nernst [110] for electrostriction; this name has been introduced for the contraction of an electrolyte solution due to the electric fields impressed on the solvent by the ionic charges. In this case the whole ionic field is indeed covered by the dielectric so that the model of the fixed capacitor is correct. For this case we may proceed with Eq.(10.72) as we did with Eq.(10.65). The results are

$$\left(\frac{\partial S}{\partial e}\right)_{T,P} = \varepsilon_o eV_c\left(\frac{\partial D}{\partial T}\right)_{e,P} \tag{10.73}$$

$$\left(\frac{\partial V}{\partial e}\right)_{T,P} = -\varepsilon_o eV_c\left(\frac{\partial D}{\partial P}\right)_{e,T} \ . \tag{10.74}$$

The volume variation in Eq.(10.68) as well as in Eq.(10.69) is opposite to the variation of the dielectric constant. The dielectric constant increases with decreasing temperature and increasing pressure; the volume changes in the opposite directions. For a comparison of Eqs.(10.69) and (10.74) we write

$$\frac{\partial[(D-1)V]}{\partial P} = V\left(\frac{\partial D}{\partial P} + \frac{D-1}{V} \cdot \frac{\partial V}{\partial P}\right) = VY \tag{10.75}$$

and compute $\partial D/\partial P$ and the coefficient Y (Table 10.1) for water at three temperatures. The data for $\partial D/\partial P$ have been taken from Owen et al. [111],

Table 10.1
Electrostriction Coefficients of Water

t	°C	0	25	70
$\frac{\partial \ln D}{\partial P}$	atm^{-1}	45.75×10^{-6}	47.72×10^{-6}	53.12×10^{-6}
$\frac{\partial \ln V}{\partial P}$	atm^{-1}	-51.66×10^{-6}	-45.84×10^{-6}	-46.75×10^{-6}
D		87.896	78.358	63.776
$\frac{D-1}{V}\cdot\frac{\partial V}{\partial P}$	atm^{-1}	-4.489×10^{-3}	-3.546×10^{-3}	-2.935×10^{-3}
$\frac{\partial D}{\partial P}$	atm^{-1}	4.020×10^{-3}	3.739×10^{-3}	3.388×10^{-3}
Y	atm^{-1}	-0.469×10^{-3}	0.193×10^{-3}	0.453×10^{-3}

those for $\partial V/\partial P$ from Kell [112]. One realizes that the pressure dependences of D and V approximately cancel each other and that the coefficient Y in Eq. (10.75) is therefore much smaller than the corresponding term $\partial D/\partial P$ in Eq. (10.74).

Regarding the temperature dependence, one will expect that the volume term as a rule will play a lesser role for polar substances, whose dielectric constants vary appreciably with temperature. For nonpolar substances the dielectric constant changes little and the expansibility term may become relatively more important.

10.4 MAGNETOSTATICS

Electromagnetic theory has developed as a magnificent example of a deductive theory that derives an immense variety of phenomena from very few basic statements. But it appears that nobody has directly and clearly discussed the pertinent generalized coordinate and force. This defect, indeed, can be perceived as the real root of Guggenheim's critical discussions [113].

In the following section, magnetization and magnetic field strength will be introduced as a generalized coordinate and the conjugate force. The discussion will proceed in principle along the same lines as the introduction of coordinates and forces in the first chapter or in Section 10.1. In other words, we extract from a few basic observations the quantities required to describe the phenomena. But the discussion will be less direct than in earlier examples. The behavior of a permanent magnet in the magnetic field of the earth will be a temporary vehicle for introducing the basic concepts.

10.41 Magnetic Moment and Field Strength.

As we did in Chapter 1, we start again from a simple experimental observation. Figure 10.5 shows a magnetic needle suspended in O by a quartz filament. The needle is not freely rotating as is a compass needle, but subject to the torque of the filament. We adjust the zero position (that would be occupied by a non-magnetic needle) arbitrarily at an angle ω from the north direction. A magnetic needle then finds an equilibrium position, say at an angle ϕ of deflection from north. A few systematic experiments with several needles and filaments show then that equilibrium depends on a quality M'' of the needle and the moment of torsion τ of the filament according to the condition

$$M'' \sin \phi = \tau(\omega - \phi) . \qquad (10.76)$$

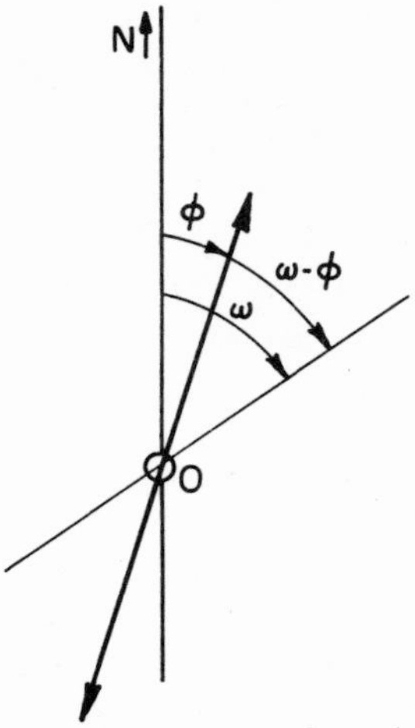

Fig. 10.5. Magnetic needle (suspended in O;ϕ angle of deflection from the north direction;ω angle between north and the zero direction of the suspension)

Moreover, experiments with the same needle and filament in various places of the earth show that M'' varies, but can be split into two factors

$$M'' = HM' \tag{10.77}$$

and that one factor, H, changes its value (slightly) from place to place; it is called the (horizontal component of the) earth-magnetic field. The other factor, M', is independent of the location and a property of the needle. We could now establish a certain magnet as a unit of the magnetic moment and determine other moments as multiples and fractions of the unit in the same way we have established a system of weights or a system of masses. Or better, we could calibrate moments and field strengths in mechanical units (centimeter-gram-second or meter-kilogram-second) by applying Eqs.(10.76) and (10.77).

The magnetic field strength can then be determined as the rotational moment produced by a magnet of unit moment.

At the same time we note that the moment of several identical magnets that are assembled all in the same direction (side-by-side or end-to-end) is the sum of the moments of the constituent parts. Therefore it makes sense to introduce the magnetic moment per unit volume, which is called the magnetization M. The molal magnetization of a substance with the molal volume $\bar{V}$ is then $M\bar{V}$.

The permanent magnet is an idealized object though of considerable technical importance and closely approached in recently developed processes.

It does not present problems of immediate thermodynamic interest.

In general, the magnetization of a substance depends on the magnetic field strength, the temperature and pressure and, moreover, on the preceding history of the object. But an equilibrium state can be attained only if the magnetization is completely determined by the variables of state. We shall therefore exclude any object in which magnetization has remained from previous treatment ("remanent magnetism") and restrict ourselves to objects whose magnetization is a well-defined function of the magnetic field strength H, the temperature and the pressure. Thus permanent magnets are excluded.

Anisotropic substances, as for instance some crystals, may respond to a magnetic field by a magnetization in a direction different from that of the field. We exclude this case, which is more complicated without contributing any essential aspect. Thus we save the trouble of using vector language, indispensable in general electromagnetism.

Concerning the magnetic field, we shall drop the restriction to the earth-magnetic field, which has been used as an illustrative example, and notice only that fields can be produced by various means (permanent magnets, currents).

Magnetization is a property of a substance; it can be varied by applying a magnetic field. Let us compare a magnetized substance with the spiral spring discussed in Section 1.24 and shown in Fig. 1.1. The corresponding properties are the length of the spring and magnetization; the one characterizes the state of the object in certain mechanical processes, the other in magnetic changes. Within certain limits we can produce according to our pleasure any length of the spring by applying a suitable force to it (by means of a weight). In the same way we can produce any amount of magnetization (within proper limits) by applying a suitable magnetic field strength H to the substance.

The two cases are not just linked by analogy. There is an essential identity: We change a property by means of applying an external force. In both cases the property is a generalized coordinate changed by means of a generalized force. We conclude that the magnetization M is a generalized coordinate and the magnetic field strength H is the conjugate generalized force.

The work required to change the magnetization is therefore HdM. We introduce a constant coefficient μ_0, the permeability of the empty space, to allow for various measuring systems. The work required to change the magnetic moment of an object occupying the volume V is now

$$dW = \mu_0 H d(VM) . \qquad (10.78)$$

This relation expresses the same statement as (10.64) for the electrostatic case.

10.42 Magnetic Properties. The ratio

$$\chi = M/H \tag{10.79}$$

is called susceptibility or volume susceptibility. If remanent magnetism is, as always here, excluded, the susceptibility is a function of temperature, pressure and field strength. At very high levels of the field strength, M approaches a saturation value so that the susceptibility decreases with increasing H; but in moderate and low fields it assumes a constant value.

The great variety in the magnetic behavior of different substances is reflected in the values of the susceptibilities of three characteristically different classes.

For a majority of substances the susceptibility has a negative small value, of the order of magnitude 10^{-5}. These substances are called diamagnetic. The variation of the susceptibility with temperature is small.

For a large class of substances, called paramagnetic, the susceptibility is positive and often appreciably dependent on the temperature. Curie's Law

$$V\chi = C_M/T \tag{10.80}$$

for temperatures below C_M is often a good approximation. The magnetic behavior of paramagnetic substances is similar to the electric behavior of dielectrics.

Very high values (sometimes as high as 10^4) are characteristic for the susceptibility of ferromagnetic substances. They often show remanent magnetism.

The correspondence of magnetic and electric terms is shown in Table 10.2. The permeability

$$\mu = \mu_o(1 + \chi) \tag{10.81}$$

and magnetic induction

$$B = \mu_o(H+M) = \mu_o(1+\chi)H = \mu H \tag{10.82}$$

are added in the table although they will not appear in the following discussion.

For various reasons the question of electromagnetic units is still in bad shape.

Maxwell's theory requires that the product of permittivity and permeability in the empty space is equal to the reciprocal square of the velocity of light in the vacuum, or

$$\varepsilon_o\mu_o = 1/(2.9979250\times10^{10})^2 = 1.112650\times10^{-20}\ \text{cm}^{-2}\text{sec}^2 \ . \tag{10.83}$$

Table 10.2
Electric and Magnetic Terms

Electric Term	Notation: This Book	Notation: Other Authors
Electric Charge	Q	
Voltage	E	ϕ, V
Charge density or dielectric displacement	q	D
Electric field strength	e	E, 4πE
Permittivity	ε	
Dielectric constant	D	ε
(Electric susceptibility)	D - 1	
Polarization	(D-1)e	P, 4πP

Magnetic Term	Notation: This Book	Notation: Other Authors
Magnetic induction	B	
Magnetic field strength	H	4πH
Permeability	μ	
Relative permeability	μ/μ_0	
Susceptibility	χ	κ, $4\pi\kappa$
Magnetization	M	J, 4πJ

One of the two constants can be chosen arbitrarily. Accordingly the units used in electrostatics, ESU, were based on $\varepsilon_0 = 1$, and those in electromagnetics, EMU, on $\mu_0 = 1$. In both systems the factor 4π appears in various general relations without manifest intrinsic reason (the entailing change in definitions is indicated in Table 10.2 in the notation column). For this reason, the two older ("unrationalized") systems are more and more replaced by a rational system in which

$$\mu_0 = 4\pi\times10^{-7}\ \mathrm{J\ s^2C^{-2}m^{-1}} = 1.2566\times10^{-7}\ \mathrm{joule\cdot sec^2\cdot Coulomb^{-2}\cdot meter^{-1}}\ . \quad (10.84)$$

A rational system is used in this chapter while the customary ESU and EMU, both in the CGS system, have been used in other places. The rational system appears generally, though not necessarily, combined with the Meter-Kilogram-Second System (MKS) or International System (SI), which is based on the units meter, kilogram, second, ampere, kelvin (temperature) and candela (luminous intensity). Numerical values and literature references can be found in a report by Mechtly [114]; general discussions of the electromagnetic units are presented in Guggenheim's book [113] and in a study by McCaig [115].

For practical application a careful study of the required units is advisable.

10.43 Molal Properties. After these circumstantial discussions the concrete results of magnetothermodynamics can be obtained without any effort. We have simply to repeat the derivation running from Eqs.(10.64) to (10.70) after replacing e, ε_o, and D-1 by H, μ_o and χ, as indicated in Table 10.2. Since relation (10.64) indeed corresponds to Eq.(10.78) in the magnetic case, we find instead of Eqs.(10.68), (10.69), (10.70)

$$\left(\frac{\partial S}{\partial H}\right)_{T,P} = \mu_o H\left(\frac{\partial(\chi V)}{\partial T}\right)_{H,P} \tag{10.85}$$

$$\left(\frac{\partial V}{\partial H}\right)_{T,P} = -\mu_o H\left(\frac{\partial(\chi V)}{\partial P}\right)_{H,T} \tag{10.86}$$

$$\left(\frac{\partial C_P}{\partial H}\right)_{T,P} = \mu_o HT\left(\frac{\partial^2(\chi V)}{\partial T^2}\right)_{H,P} . \tag{10.87}$$

These relations are particularly important at very low temperatures. Here they can be simplified for liquids and solids since the derivatives of the volume with respect to temperature and pressure approach zero. Moreover, Curie's Law (Eq.(10.80)) is also often applicable at low temperatures.

10.5 RADIATION

In daily life we encounter radiation primarily as a form of energy transmission. Radiation is emitted by material objects, transmitted through space and absorbed by other objects. In the following we exclude specific radiations produced selectively at certain frequencies by special reactions and at the passage of electricity through a neon tube.

Since radiation propagates at a finite velocity c, there is a certain amount of energy in a finite space enclosed by emitting and absorbing or reflecting walls. If the emitting and absorbing materials are kept at a definite temperature, an equilibrium between radiation and walls is established. We call the energy per unit volume u. Since it depends on the temperature, it is associated with a finite amount of entropy s per unit volume. A very small but finite pressure P is exerted by radiation.

10.51 Radiation Pressure. The derivation in Chapter 3 of the pressure of a perfect gas

$$P = \frac{vN_L}{6V} \cdot \frac{2\bar{M}v}{N_L} \qquad (3.19)$$

though primitive can be immediately applied to the pressure due to radiation. We substitute n for the number N_L/V of radiation units in the unit volume, and p for the linear momentum $\bar{M}v/N_L$ of each unit. The pressure is therefore given by

$$P = nvp/3 \; . \qquad (10.88)$$

For the perfect gas we could replace the momentum by the mass m and the velocity v of a molecule so that

$$P = nmv^2/3 \qquad (10.89)$$

or equal to 2/3 of the kinetic energy per unit volume.

According to Maxwell's theory the pressure can be expressed by the energy u per unit volume as

$$P = u/3 \; . \qquad (10.90)$$

Boltzmann [116] mentioned that a theory of radiation by particles would lead to a pressure twice as high as Eq.(10.90), and Planck [117] presented a careful and detailed derivation, concluding that the radiation pressure makes a decisive distinction between the particle and wave theories of radiation.

Both consider the total radiation energy to be kinetic energy. But if we may attribute to radiation a kinetic and a potential energy in the same manner we assign them to any harmonic oscillator, the radiation pressure is in perfect accord with the fact that the total energy of a harmonic oscillator is twice its average kinetic energy.

The distinction between particles and waves, however, becomes irrelevant in quantum mechanics. Here the energy of a photon is assumed to be given by the product of the frequency ν and Planck's constant h, and the linear momentum of a photon is assumed to be the energy divided by its velocity c. According to Eq.(10.88) the contribution of the photons of the frequency ν_i to the pressure is therefore

$$P_i = n_i c \cdot h\nu_i/(3c) = n_i h\nu_i/3 \qquad (10.91)$$

or equal to one third of the energy contribution of the same frequency. The summation over all frequencies furnishes Maxwell's relation (10.90).

10.52 Stefan-Boltzmann's Law. The only kind of work we have to consider for radiation in equilibrium at the temperature T is the work $-PdV$ of the radiation pressure. Formally the problem is the same as the behavior of a pure phase (Eq.(3.1)). We may therefore express relation (2.32) as

$$d(uV) = TdS - PdV \tag{10.92}$$

replacing the total energy by the energy density u. We need two independent variables and choose T and V.

From

$$dS = [d(uV) + PdV]/T \tag{10.93}$$

and Eq.(10.90) we obtain

$$dS = (Vdu + \tfrac{4}{3}udV)/T \tag{10.94}$$

so that

$$\left(\frac{\partial S}{\partial T}\right)_V = \frac{V}{T}\frac{du}{dT}; \qquad \left(\frac{\partial S}{\partial V}\right)_T = \frac{4u}{3T} \tag{10.95}$$

since u depends only on T but not on V. Cross differentiation furnishes

$$\frac{\partial^2 S}{\partial T \partial V} = \frac{1}{T}\frac{du}{dT} = \frac{4}{3T}\frac{du}{dT} - \frac{4u}{3T^2} \tag{10.96}$$

or

$$\frac{1}{u}\,du = \frac{4}{T}\,dT \tag{10.97}$$

$$u = aT^4 \; . \tag{10.98}$$

This relation, empirically found by Stefan, has been derived by Boltzmann.

The entropy of equilibrium radiation is according to Eq.(10.95) given by

$$\left(\frac{\partial S}{\partial T}\right)_V = \frac{V}{T}\cdot 4aT^3 = 4aVT^2 \tag{10.99}$$

and the entropy density is

$$s = S/V = 4aT^3/3 \; . \tag{10.100}$$

Debye's function (4.55) for the heat capacity of a crystal at low temperatures leads to a similar relation

$$S = \int_0^T (C_V/T)dT = DT^3/3 \; . \tag{10.101}$$

Although there is no real link between the two relations, we may note that both apply to vibrations and that any temperature is a low temperature for a photon.

The energy density and the pressure of radiation are extremely small under ordinary conditions. The coefficient of Eq.(10.98) is

$$\begin{aligned} a &= 7.56471\times10^{-16}\ \text{joule}\cdot\text{meter}^{-3}\text{kelvin}^{-4} \\ &= (7.56471\pm0.00129)\cdot10^{-15}\ \text{erg}\cdot\text{cm}^{-3}\text{deg}^{-4}. \end{aligned} \tag{10.102}$$

At 298 K the radiation pressure in equilibrium amounts therefore to

$$P = 2.00\times10^{-5}\ \mathrm{erg\cdot cm^{-3}} = 1.98\times10^{-11}\ \mathrm{atm}\ . \qquad (10.103)$$

But it increases so rapidly with temperature that it becomes important in the interior of stars.

SUMMARY OF THE TENTH CHAPTER

The application of the laws of thermodynamics to a wide variety of phenomena illuminates the flexibility of thermodynamic methods. Sometimes thermodynamic discussions appear to be collections of different rules arbitrarily established for this case or that one. But in this chapter, and indeed in this whole book, we have tried to show clearly the strong structure of thermodynamics and the firm linkage of the most widely varied applications with this structure. "Unity in variety" is achieved in thermodynamics by the search for generalized coordinates and forces.

REFERENCES

(101) E. A. Guggenheim, J. Chem. Phys., 13 (1945) 253.
(102) I. Langmuir, J. Am. Chem. Soc., 40 (1918) 1368.
(103) O. Redlich and D. L. Peterson, J. Phys. Chem., 63 (1959) 1024.
(104) S. Brunauer, P. H. Emmett and E. Teller, J. Amer. Chem. Soc., 60 (1938) 309.
(105) D. W. Breck, W. G. Eversole, R. M. Milton, T. B. Reed, T. L. Thomas, J. Amer. Chem. Soc., 78 (1956) 5963.
(106) D. L. Peterson and O. Redlich, J. Chem. Eng. Data, 7 (1962) 570.
(107) A. J. P. Martin and R. L. M. Synge, Biochem. J., 35 (1941) 1358. G. Baylé and A. Klinkenberg, Rec. Trav. Chim., 76 (1957) 593.
(108) H. K. Schachman, Ultracentrifugation in Biochemistry, Academic Press, New York (1959).
(109) T. F. Young, K. A. Kraus and J. S. Johnson, J. Chem. Phys., 22 (1954) 878; J. Am. Chem. Soc., 76 (1954) 1436.
(109.1) H. S. Frank, J. Chem. Phys., 23 (1955) 2023.
(110) P. Drude and W. Nerst, Z. physik. Chem., 15 (1894) 79.
(111) Owen, Miller, Milner and Cogan, J. Phys. Chem., 65 (1961) 2065.
(112) G. S. Kell and E. Whalley, Roy Soc. Phil. Trans., A 258 (1965) 565.

(113) E. A. Guggenheim, Proc. Roy. Soc., A 155 (1936) 49, 70; Thermodynamics, 3rd ed., Interscience Publ. Inc., New York, 1957.

(114) E. A. Mechtly, National Aeronautics and Space Administration, NASA SP-7012 (1969).

(115) M. McCaig, Chapter 2 in "Permanent Magnets and Magnetism" ed. by D. Hadfield, John Wiley, New York, 1962.

(116) L. Boltzmann, Ann. Phys., 258 (1884) 291.

(117) M. Planck, Theorie der Wärmestrahlung, 2nd ed., J. A. Barth, Leipzig, 1913. Translated by M. Masins, Dover Publ., New York, 1959.

APPENDIX 11. NOTES TO STATISTICAL THERMODYNAMICS

The following comments are intended to remove a few minor obstacles to the understanding of methods applied in molecular theory. They are neither rigorous nor profound discussions.

11.1 COMBINED PROBABILITY

Whenever we introduce probability we assume a model. The logical consequences of the model are one thing, the reasonableness of assuming the model is another thing.

As a simple example we take two dice, which we assume to be "true," i.e., we assume that after a sufficient number of throws the frequency of appearance of any specified number will be one sixth of the total number of throws. We also assume that they do not influence each other (which would not be true, for instance, if they are magnetized). Thus we assume that the 36 possible results shown in Table 11.1 are equally probable.

Table 11.1
Two Dice

11	21	31	41	51	61
12	22	32	42	52	62
13	23	33	43	53	63
14	24	34	44	54	64
15	25	35	45	55	65
16	26	36	46	56	66

With this assumption, all questions can be answered simply by counting. For instance, the probability W_1 of obtaining a three with the first die is $6/36 = 1/6$, since six out of thirty-six numbers in the table show 3 in the first place. The probability W_2 of obtaining a four with the second die is also $1/6$. The probability of having 34 is one sixth out of one sixth or

$W_1 \cdot W_2 = 1/36$. It is readily seen that combination of probabilities of independent events always requires multiplication, thus leading to Eqs.(4.1) and (4.3).

11.2 PROBABILITY AND ENTROPY

It is easily seen that the logarithm function satisfied Eq.(4.4), which expresses the fact that the entropy combines by addition while the probability combines by multiplication. It can be shown that no other relationship satisfies this condition.

We consider a system of two independent objects as we did in Section 4.1. We change the state of the first object, leaving the second unchanged. Then we derive from Eq.(4.4)

$$\left(\frac{\delta S(W)}{\delta W_1}\right)_{W_2} = \frac{dS(W)}{dW} \cdot \left(\frac{\delta W}{\delta W_1}\right)_{W_2} = \frac{dS(W_1)}{dW_1} \quad . \tag{11.1}$$

The differentiation is indicated by an ordinary d when the function depends only on a single variable.

From Eq.(4.3) we derive

$$\left(\frac{\delta W}{\delta W_1}\right)_{W_2} = W_2 \tag{11.2}$$

so that

$$\frac{dS(W_1)}{dW_1} = \frac{dS(W)}{dW} \cdot W_2 \tag{11.3}$$

or, after multiplication by W_1,

$$W_1 \frac{dS(W_1)}{dW_1} = W \frac{dS(W)}{dW} \quad . \tag{11.4}$$

In other words, the value of Eq.(11.4) is always the same, whatever the value of the argument W of the function S(W) may be. This conclusion is general since we can vary the first object in an entirely arbitrary manner. Therefore Eq.(11.4) must be a constant k

$$W \frac{dS(W)}{dW} = k \tag{11.4}$$

which leads necessarily to Boltzmann's theory (4.6).

11.3 STIRLING'S APPROXIMATION

The approximation (4.16) frequently used for N! can be made plausible by a comparison of Fig. 11.1 and Fig. 11.2. The area under the curve of

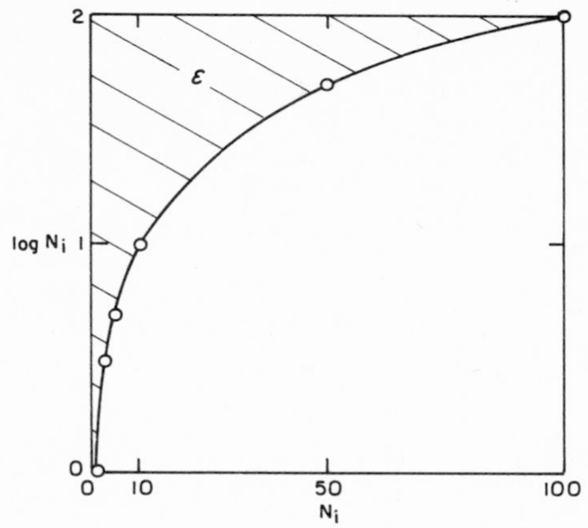

Fig. 11.1. The Sum of log N_i

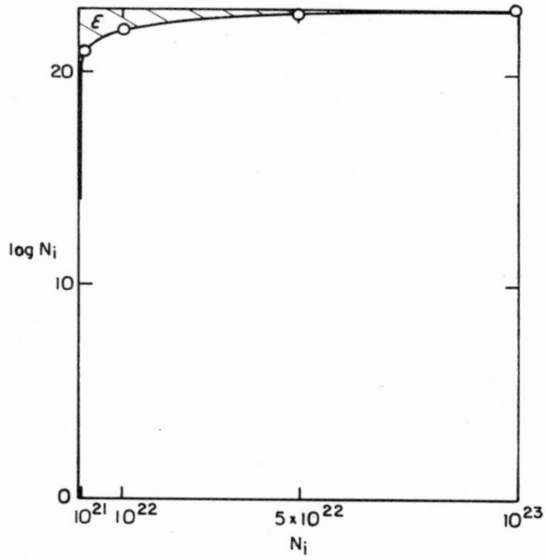

Fig. 11.2. Stirling's Approximation

Fig. 11.1 represents

$$\sum_{i=1}^{100} \log N_i = \log(100!). \tag{11.5}$$

In Fig. 11.2 we see, on a different scale, the area representing

$$\sum_{i=1}^{10^{23}} \log N_i = \log(10^{23}!) . \tag{11.6}$$

Stirling's approximation N log N is represented by the square in 11.1 and the rectangle in 11.2. It is immediately apparent that with increasing N the error ε decreases rapidly in comparison to the total area.

Stirling's formula furnishes also an approximation of the error ε. In the applications required in Chapter 4 the terms containing ε cancel.

11.4 THE PARTITION FUNCTION

The partitition function (German: Zustandssumme) introduced in Eq.(4.29) is the principal tool for applying molecular theory in thermodynamics. Its basis is given by Eq.(4.23)

$$\ell n\ W = N_L\ \ell n\ N_L - \sum g_j N_j\ \ell n\ N_j \qquad (11.7)$$

which describes the probability of a certain distribution of N_L molecules in states of different energies ε_j. The number of molecules in the state j is N_j, and the number g_j of different states with the same energy ε_j is indicated by the statistical weight g_j.

The problem is now to find that distribution N_1, N_2, ... N_j, ... for which the probability W is a maximum, with the restrictions that the total number N_L of the molecules and their total energy $\bar{E}$ are prescribed. These restrictions are expressed in Eqs.(4.24) and (4.25), namely,

$$\bar{E} = \sum g_j\ \varepsilon_j\ N_j \qquad (11.8)$$

$$N_L = \sum g_j\ N_j\ . \qquad (11.9)$$

The requirement that W or ℓn W have a maximum value implies that the change $\delta \ell n$ W produced by any possible (small) changes δN_1, δN_2, ... δN_j, ... is zero. But not all of the variations δN_j can be arbitrarily chosen since they are restricted by

$$\delta E = \sum g_j\ \varepsilon_j\ \delta N_j = 0 \qquad (11.10)$$

$$\delta N_L = \sum g_j\ \delta N_j = 0\ . \qquad (11.11)$$

We can, for instance, solve the two equations

$$\varepsilon_o\ \delta\ N_o + g_1\ \varepsilon_1\ N_1 = -\sum_{j=2} g_j\ \varepsilon_j\ \delta N_j \qquad (11.12)$$

$$\delta N_o + g_1\ \delta N_1 = -\sum_{j=2} g_j\ \delta N_j \qquad (11.13)$$

for δN_o and δN_1; then the variations δN_2, δN_3 ... are independent of one another. The maximum for W is now obtained by the condition

$$\delta\ \ell n\ W = -\sum g_j(\ell n\ N_j + 1)\ \delta N_j \qquad (11.14)$$

where δN_0 and δN_1 are replaced by δN_2, δN_3 ... according to the solution of Eqs.(11.12) and (11.13).

The procedure of Lagrange is equivalent but more elegant. We introduce two constants α and β with the proviso that we shall later determine these quantities in such a manner that Eqs.(11.8) and (11.9) will be satisfied. We multiply Eq.(11.10) by $-\beta$ and Eq.(11.11) by $1+\alpha$ and add both to Eq.(11.14). This furnishes

$$\delta \ell n\, W + (\alpha + 1)\delta N_L - \beta\delta\bar{E} = -\sum g_j(\ell n\, N_j - \alpha + \beta\varepsilon_j)\delta N_j = 0. \quad (11.15)$$

But now we need not replace δN_0 and δN_1 by Eqs.(11.12) and (11.13) because these conditions or the equivalent Eqs.(11.8) and (11.9) will be satisfied in any case by the choice of α and β. Therefore the condition for the maximum is now that Eq.(11.15) equals zero for any arbitrary choice of δN_0, δN_1, ... δN_j, ... This means that the factors of <u>all</u> δN_j's must be zero, or

$$\ell n\, N_j = \alpha - \beta\varepsilon_j; \qquad N_j = e^{\alpha} \cdot e^{-\beta\varepsilon_j}. \quad (11.16)$$

Since

$$N_0 = e^{\alpha} \cdot e^{-\beta\varepsilon_0} \quad (11.17)$$

we eliminate α by means of N_0

$$N_j = N_0\, e^{-\beta(\varepsilon_j-\varepsilon_0)} \quad (11.18)$$

so that according to Eq.(11.9)

$$N_L = N_0(1 + g_1\, e^{-\beta(\varepsilon_1-\varepsilon_0)} + \ldots + g_j\, e^{-\beta(\varepsilon_j-\varepsilon_0)} + \ldots) = N_0Q. \quad (11.19)$$

The partition function

$$Q = \sum g_j\, e^{-\beta(\varepsilon_j-\varepsilon_0)} \quad (11.20)$$

can now replace N_0 in Eq.(11.18) so that

$$N_j = (N_L/Q)\, e^{-\beta(\varepsilon_j-\varepsilon_0)}. \quad (11.21)$$

In order to satisfy condition (11.8) we differentiate Eq.(11.20) with respect to β

$$dQ/d\beta = -\sum g_j(\varepsilon_j-\varepsilon_0)\, e^{-\beta(\varepsilon_j-\varepsilon_0)}. \quad (11.22)$$

Introducing

$$\bar{E}_0 = \varepsilon_0\, N_L \quad (11.23)$$

we find from Eqs.(11.8), (11.9) and (11.21)

$$\bar{E} - \bar{E}_0 = \sum g_j(\varepsilon_j-\varepsilon_0)N_j = (N_L/Q)\sum g_j(\varepsilon_j-\varepsilon_0)\, e^{-\beta(\varepsilon_j-\varepsilon_0)}. \quad (11.24)$$

Comparison of Eqs.(11.22) and (11.24) shows

$$dQ/d\beta = -(Q/N_L)(\bar{E} - \bar{E}_0). \quad (11.25)$$

Furthermore we find from Eqs.(11.7) and (11.18)

$$\ell n\ W = N_L\ \ell n\ N_L - g_j\ N_j[\ell n\ N_0 - \beta(\varepsilon_j - \varepsilon_0)] \tag{11.26}$$

or, according to Eqs.(11.9) and (11.24),

$$\ell n\ W = N_L\ \ell n(N_L/N_0) + \beta(\bar{E} - \bar{E}_0) \tag{11.27}$$

According to Eq.(11.19) we replace N_L/N_0 by Q and obtain the molal entropy from Eq.(4.6) as

$$\bar{S} - \bar{S}_0 = kN_L\ \ell n\ Q + k\beta(\bar{E} - \bar{E}_0)\ . \tag{11.28}$$

Differentiation leads to

$$\frac{dS}{d\beta} = \frac{kN_L}{Q} \cdot \frac{dQ}{d\beta} + k(\bar{E} - \bar{E}_0) + k\beta \frac{d\bar{E}}{d\beta}\ , \tag{11.29}$$

or, in view of Eq.(11.25), to

$$\frac{dS}{d\beta} = k\beta \frac{d\bar{E}}{d\beta} \tag{11.30}$$

so that

$$\frac{d\bar{S}}{d\bar{E}} = k\beta\ . \tag{11.31}$$

The comparison with the definition of the entropy (2.18) shows that

$$\beta = \frac{1}{kT}\ . \tag{11.32}$$

This is the condition that satisfied Eq.(11.8).

Boltzmann's law (4.26) is now expressed by introducing Eq.(11.32) into Eq.(11.18).

11.5 AN EXAMPLE OF MOLECULAR VIBRATIONS

The basic features of molecular vibrations have been briefly discussed in Section 4.4. Raman and infrared spectra constitute the primary material from which the various kinds of vibrations are deduced. The first step, the interpretation of the observed frequencies, is based on plain mechanics; the second step, the theory of line or band intensities and selection rules, is based on quantum mechanics and group theory. The assignment of frequencies to the theoretical vibration forms is deduced from intensities and polarization of the lines, and on comparison of isotopic molecules and similar molecules.

Water and carbon dioxide have been chosen as examples for vibration forms in Fig. 11.3 because the comparison illustrates a degenerate state.

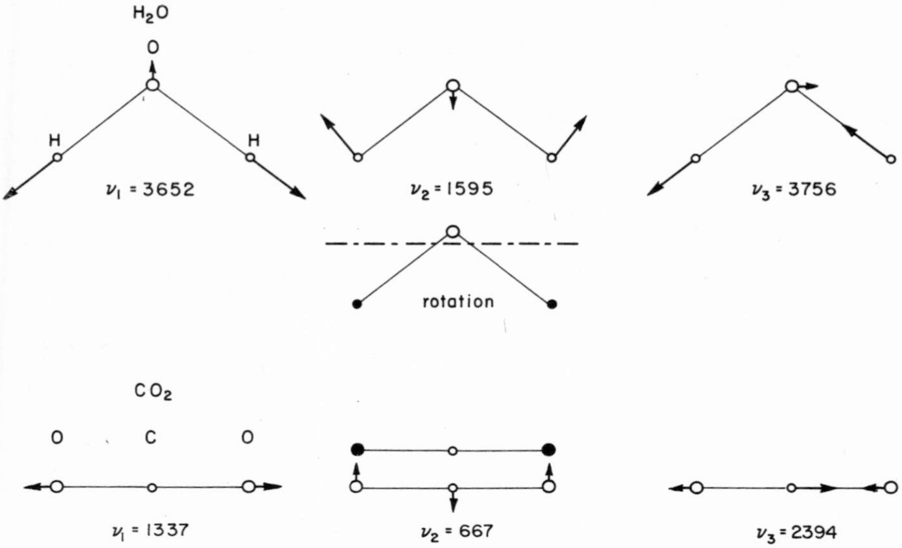

Fig. 11.3. Vibration of H_2O and CO_2 (the wave numbers of the vibrations are given in cm^{-1})

The two molecules differ in the angle between the valences and the masses of the atoms. The diagram is not in true scale with respect to the sizes and distances of the atoms or the amplitudes (the sizes of the atoms would be much larger, the amplitudes smaller). If we imagine a continuous transformation from H_2O to CO_2 the angle between the valences would increase and the amplitudes of the outer atoms decrease. Each vibration of water transforms into the one beneath it.

But according to Table 4.2 carbon dioxide has 4 vibrations (and two rotations because there is no rotation excited around the main axis of the molecule). Water has 3 vibrations and the regular three rotations, one of which proceeds around the axis indicated in Fig. 11.3 by a broken line. It is this rotation which transforms, together with the second water vibration, into the second vibration of CO_2. The water rotation, indeed, can be represented by motions perpendicular to the molecular plane (O up, H down); this kind of motion becomes the second CO_2 vibration turned around the main axis by 90°. This kind of vibration takes place of course in every direction but its representation requires two independent coordinates (in the plane of the diagram and perpendicular to it). The restoring forces do not depend on the azimuth of the vibration and the energies of the two independent vibrations are equal.

In computing the vibrational contributions to the heat capacity of carbon dioxide, one has to count the contribution of ν_2 twice.

APPENDIX 12. COMPUTATION METHODS

The voyage from the basic theory to practical applications is properly concluded with a brief discussion of efficient computing. There are indeed a few simple rules which may be helpful for everyone; they are important in preparing a scheme for routine computations.

12.1 TABULAR COMPUTATION

A series of computations of the same kind is always carried out in a table similar to Table 9.2, which represents a complete computation table.

It is indeed an important point to perform all computations in a simple table. One assigns a column to each specific problem and a line to each computation step. The advantages are quicker and more reliable performance in the repetition of the same step for the whole line, and a very good chance to detect errors either right in the operation or by later inspection. The inverse assignment (a line to each problem and a column to each computation step)is almost always less convenient (numerical integration may sometimes be an exception).

Careful preparation is required for the resolution of a calculation into elementary steps which can be performed each in one line. It pays to assemble first all algebraic relations and then all numerical data including conversion factors to be used on a sheet. Then the simple computation steps are indicated in the first column of the computation table.

The use of separate slips of paper for side calculations is more than anything else liable to mess up a computation.

It is worth while to give a lot of thought to the final representation of the result. In Table 9.5, for example, the result is essentially given by the residue r. But for a satisfactory discussion one also needs the mole fraction of SO_2 and the amount of pollution. Often the best representation of the results will be graphical;in this case the preparation of the data for the diagram should be included in the main computation.

12.2 DIAGRAMS

The advantages of a diagram for quickly conveying a wealth of information and as a means of comparison are obvious.

There are a few practical rules. It helps if we restrict ourselves to millimeter-cross section paper and introduce only 1, 2 or 5 as scale factors; thus we acquire quickly adequate practice both in drawing diagrams and in reading results from them. With moderate care, data can be safely recovered within 0.5 mm; a careful worker can do a little better.

Function scales such as logarithmic scales or paper for plotting K and T as log K and 1/T are sometimes convenient but entail some loss in precision by interpolation errors.

The limitations of the graphical method are felt in averaging and smoothing experimental data, in interpolation and extrapolation, and in other computational applications. Cross section paper is available in 500 mm wide rolls; a diagram as large as 500×500 mm is slightly inconvenient and nothing at all would be gained by further expansion. The error of direct graphical representation therefore cannot be reduced to less than 0.1% of the total range of a variable.

The remedy is the method of deviation functions, developed by Lewis and Randall. One chooses a simple approximation function and plots the difference between the experimental results and the selected function. For example, instead of log γ' in Fig. 8.5, we may plot the deviation function

$$\log \gamma'' = \log \gamma' + 0.511\, m^{0.5} \qquad (12.1)$$

which is obtained by subtracting the approximation of Debye and Hückel (8.43) from the experimental data. The deviation function is plotted against m rather than $m^{0.5}$ because in a power series of log γ the first deviations from Debye and Hückel are proportional to m. The result is shown in Fig. 12.1.

The diagram shows the scattering a high dilution. There is a systematic deviation increasing with increasing concentration. The straight line

$$\log \gamma'' = -1.8790 + 0.755m \qquad (12.2)$$

satisfactorily represents the data for low concentrations but is not necessarily the best expression of the data. For the best representation one would have to include the observations at higher concentrations; the addition of a term with $m^{1.5}$ or m^2 may lead to a small change of the coefficients of (12.2).

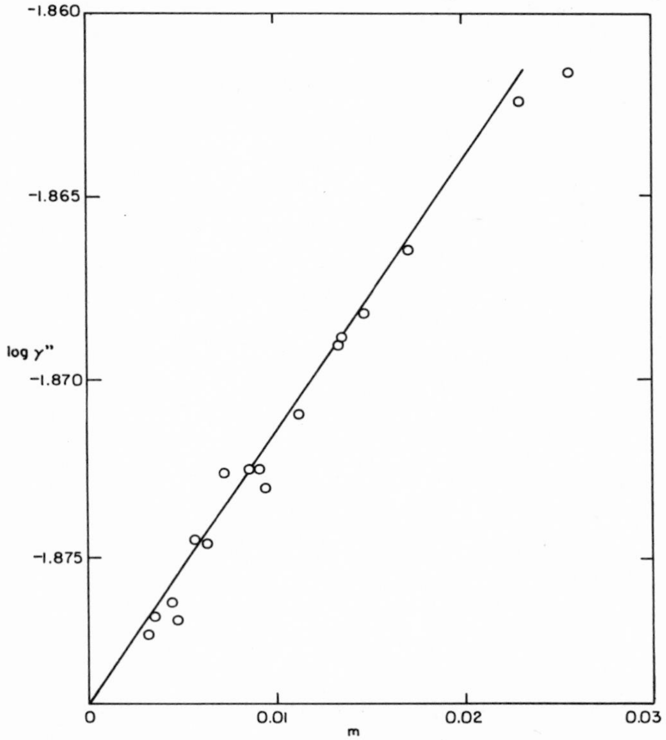

Fig. 12.1. Deviation Function for the Activity Coefficient

For the small range up to m = 0.026 of Fig. 12.1 we have easily gained an advantage over Fig. 8.5 by adjusting the scale factors to the new range of the variables. But we lose this advantage at higher concentrations because of the increase in the range of deviations from the limiting law. For the whole range we would abandon the recourse to the theoretical approximation and plot the deviation function

$$\log \gamma''' = \log \gamma' + 0.300 m^{0.5} . \qquad (12.3)$$

The experimental information is exhaustively expressed in a diagram if the directly observed data scatter noticeably. If they lie exactly on a smooth curve the design of the diagram is insufficient. This is an essential point in graphical computation. The rule does not apply, of course, to diagrams designed for information rather than computation.

From Eqs.(8.88),(12.1) and (12.2) we obtain as a preliminary representation for dilute solutions

$$\begin{aligned} \log \gamma &= \log \gamma' + 0.5\, E^o/0.059158 \\ &= 1.8790 + \log \gamma'' - 0.511 m^{0.5} \\ &= -0.511 m^{0.5} + 0.755 m \;. \end{aligned} \tag{12.4}$$

12.3 ALGEBRAIC REPRESENTATION

There has always been a competition in the representation of experimental data between tables, diagrams and algebraic equations. The relative advantages depend in various ways on the precision of the data, on the purpose of the user, and on convenience in handling the data. No general comparative discussion would be useful.

Automatic computation has noticeably shifted the balance in favor of algebraic representation. This development has accentuated the key problem, namely, the smoothing of experimental data with the intent of extracting the best information from them. The judgment of the operator drawing the curves in Figs. 8.5 or 12.1 should be replaced by objective rules. The solution of this problem is usually seen in the method of least squares; there are indeed good arguments for the adoption of this method.

For algebraic representation of the values Y_i observed for the values X_i of the independent variable one has to select a function F(X), which may be linear in X; it may be a power series of any other relation. The coefficients or parameters of this function are now the unknowns. As soon as they are determined, we have the algebraic representation of the observations.

The method of least squares aims to prescribe a unique and definite rule for the best values of the parameters appearing in the selected function F(X). If the probable errors in the observed values Y_i are equal, the sum S of the squares of the deviations

$$S = \sum_i [Y_i - F(X_i)]^2 \tag{12.5}$$

must be a minimum for the variation of any single one of the parameters. This condition furnishes the equations requisite for the computation of the parameters.

Frequently there is no reason to assume that the probable errors of the observed data differ from point to point. In this case the minimum of S furnishes indeed the best parameters. But such an assumption is often conspicuously wrong, particularly in the thermodynamics of solutions. As an example we choose the activity coefficient of the second component of a binary solution, as given by Eq.(7.11), namely,

$$\gamma_2 = Py_2/(p_2^o x_2) \tag{12.6}$$

where P is the total pressure, p_2^o is the vapor pressure and x_2, y_2 the mole fractions of the second component in the liquid and vapor. We choose as the observed quantity

$$Y = \ln \gamma_2 = \ln P + \ln y_2 - \ln p_2^o - \ln x_2 \tag{12.7}$$

and obtain for a small error ε in x_2 the variation δY in Y, namely

$$\delta Y = \varepsilon \partial Y/\partial x_2 = -\varepsilon/x_2 \ . \tag{12.8}$$

If the experimental procedure justifies the assumption that the probable error in x_2 has the same value for the whole set of measurements, then its contribution to the probable error in Y varies widely with the mole fraction and increases without a limit on approach of $x_2 = 0$. Figure 6.5 illustrates this case: The two points near $x_1 = 1$ of an obviously excellent set deviate greatly.

The method of least squares prescribes in this case weighting of the observed data. The weights w_i are to be chosen proportional to the reciprocal squares of the probable errors in Y and the sum (12.5) is to be replaced by

$$S = \sum_i w_i[Y_i - F(X_i)]^2 \ . \tag{12.9}$$

Moreover, in the estimation of the probable error in Y_i one has to consider <u>all</u> sources of error. The square of the probable error in Y_i is the sum of the squares of the contributions from each error source. Thus in the example (12.6) the weight of a single observation is

$$w_i = 1\bigg/\left[\left(\frac{\varepsilon'}{P}\right)^2 + \left(\frac{\varepsilon''}{y_2}\right)^2 + \left(\frac{\varepsilon'''}{p_2^o}\right)^2 + \left(\frac{\varepsilon^{iv}}{x_2}\right)^2\right] \tag{12.10}$$

where ε', ε'', ε''', ε^{iv} are the probable errors in the determination of P, y_2, p_2^o, x_2. The estimation of ε', ... is often a difficult problem; in particular, these quantities need not be constants for a given set of measurements. It is fairly obvious, for example, to assume a constant analytical error independent of the concentration; actually, however, the error often is smaller for low concentrations so that the strong influence of y_2 and x_2 is exaggerated if ε'' and ε^{iv} are considered to be constant.

This discussion shows that the method of least squares does not at all present an automatically correct and objective procedure. The object of the operator's judgment has changed from drawing the best curve to the balanced estimation of probable errors. There is now available a considerable number of more or less sophisticated programs for least-squares computations; they can save a lot of time and effort. But the best program cannot furnish better results than the information put in. An estimate of the probable errors for each observation is an essential part of this information.

APPENDIX 13. UNIVERSAL CONSTANTS AND OTHER NUMERICAL VALUES

Metrology, the doctrine of measurement and of the numerical representation of observations, is a fairly involved and still developing branch of science. We mention in the following only a few points, important for the use of the fundamental constants.

Strictly speaking, a numerical datum is complete only if its probable error is given. We express in the following the probable error of important data in parts per million (ppm). Conventionally one presents a figure with so many digits that the probable error amounts to several units of the last place if one does not indicate it explicitly. It is good practice, particularly in involved computations, to carry along one or two places more in the computation and to round off the result.

13.1 BASIC AND DERIVED UNITS

Every measurement is the final step in a series of comparisons between two objects which links the measured object step by step to a basic standard. The choice and realization of the standards has a long and complicated history reflected in the periodically convened General Conferences on Weights and Measures. Several important changes have been made in recent years.

As before, the standard kilogram (kg) kept at Sèvres, France, is the basic unit of mass. The use of the kilogram as a unit of force is not approved. The meter (m) is based on the wave length of a carefully specified spectral line, the second (s) on the period of vibration of another line. The ampere (A) is based on the mutual magnetic force exerted by the current in two specified conductors.

The unit of temperature is the kelvin (K) defined as 1/273.16 of the temperature at which ice, water and steam are in equilibrium. The freezing point of water at atmospheric pressure is 0.010 K lower, i.e., 273.150 K and the centigrade or Celsius temperature is $t = T - 273.150$.

As another basic unit the candela (cd) has been defined for luminous intensity.

All other units are expressed by these six basic units. As a unit of force the newton (N) has been introduced; it is that force which gives a mass of 1 kilogram an acceleration of 1 meter per second per second. The joule (J) is the work done when the point of application of 1 newton is displaced a distance of 1 meter in the direction of the force.

The watt (W), volt (V), ohm (Ω), coulomb (C), and farad (F) are in turn derived from the preceding units without any essential change.

The liter is now defined as a cubicdecimeter or 0.001 m^3.

These definitions have straightened out the previous confusing array of "international," "absolute" and various other units. The only unit based on an individual object is the kilogram. All other units can be reproduced at any time in the laboratory.

13.2 NUMERICAL VALUES

Numbers established by definition are indicated in the following by an asterisk; they are of course not subject to experimental error.

Most of the numerical values of Tables 13.2, 13.3 and 13.4 are taken or derived from E. A. Mechtly's article "The International System of Units" (NASA SP-7012, Washington, D.C., 1969).

The use of cm^{-1} as an energy unit (Table 13.3) is justified by the relations of Planck and Einstein that show proportionality of energy and wave number (expressed in cm^{-1}).

A voltage given as "electronvolt" means actually an energy, namely, the amount determined by the observed voltage and the associated charge, either the electron charge (for one molecule or another particle) or the Faraday equivalent (for a mole or an equivalent). Hence corresponding numbers in Tables 13.2 and 13.4 are equal.

The last figures in Table 13.4 give the energy set free by the disappearance of mass in a fission reaction.

Table 13.1
Prefixes

Factor	Prefix	Symbol
10^{12}	tera	T
10^{9}	giga	G
10^{6}	mega	M
10^{3}	kilo	k
10^{2}	hecto	h
10	deka	da
10^{-1}	deci	d
10^{-2}	centi	c
10^{-3}	milli	m
10^{-6}	micro	μ
10^{-9}	nano	n
10^{-12}	pico	p
10^{-15}	femto	f
10^{-18}	atto	a

Table 13.2 - See page 278

Table 13.3
Pressure Conversion

	newton m^{-2}	atm.
1 atm	$1.01325^{*}\times10^{5}$	1.00
1 mm mercury (0°C)	1.33322×10^{2}	1.31579×10^{-3}
1 cm water (4°C)	9.80638×10^{1}	9.67814×10^{-4}
1 millibar	$1.00^{*}\times10^{2}$	0.86923×10^{-4}
1 newton	1.00	9.86923×10^{-6}

Table 13.2
Universal Constants

		Unit	Error (ppm)
Gas constant R	8.31434×10^{3}	J $kmole^{-1}K^{-1}$	42
	8.31434	J $mole^{-1}K^{-1}$	
	1.98717	cal $mole^{-1}K^{-1}$	
	8.20562×10^{1}	cm^3 atm $mole^{-1}K^{-1}$	
ℓn 10	2.302585		
R ℓn 10	4.57563	cal $mole^{-1}K^{-1}$	
298.150 R ℓn 10	1.36422×10^{3}	cal $mole^{-1}$	
Standard Volume of the Ideal Gas	2.24136×10^{1}	m^3 $kmole^{-1}$	
	2.24136×10^{1}	lit $mole^{-1}$	
	2.24136×10^{4}	cm^3 $mole^{-1}$	
Faraday equivalent F	9.648670×10^{7}	C $kequ^{-1}$	5.5
	9.648670×10^{4}	C equ^{-1}	
	9.648670×10^{4}	J V^{-1} equ^{-1}	
	2.306087×10^{4}	cal V^{-1} equ^{-1}	
298.150 R ℓn 10/F	5.91575×10^{-2}	V	
Loschmidt's Number N_L per mole (Avogadro's Constant)	6.022169×10^{26}	k $mole^{-1}$	6.6
	6.022169×10^{23}	$mole^{-1}$	
Boltzmann's Constant k	1.380622×10^{-23}	J K^{-1}	43
Planck's Constant h	6.626196×10^{-34}	J s	7.6
Velocity of Light c	2.9979250×10^{8}	m s^{-1}	0.33
hc/k	1.438833	cm K	43
Electron Charge	1.6021917×10^{-19}	C	4.4
Electronvolt	1.6021917×10^{-19}	J	

Table 13.4
Energy Relations

1 J = 1 Ws = 1 VC = 1 VAs = 0.239006 cal

1 kilowatt hour = $3.600^* \times 10^6$ J = 3.600* MJ

1 cm^{-1} (see Section 4.4)	=	1.2398541×10^{-4} eV
	=	1.986484×10^{-23} J/molecule
	=	11.96294×J/mole
	=	2.859212 cal/mole
1 electronvolt (eV)	=	$1.6021917 \times 10^{-19}$ J/molecule
	=	9.648670×10^{4} J/mole
	=	2.306087×10^{4} cal/mole
	=	1 (Faraday equivalent)×V/mole
1 lit atm	=	$1.01325^* \times 10^{2}$ J
	=	24.2173 cal
1 gram/mole	=	8.987554×10^{13} J/mole
	=	2.148077×10^{13} cal/mole
	=	9.314811×10^{8} eV

PROBLEMS

It is advisable to read Appendices 12 and 13 before solving the more complicated problems proposed in the following.

The author is obliged to Professor Leo Brewer for his kind permission to use a number of problems from his lecture notes; some of them have been slightly changed for the present purpose.

Data References

LRPB: Lewis-Randall-Pitzer-Brewer, Thermodynamics, McGraw-Hill Book Co., New York, 1961.

Perry's Handbook: J. H. Perry, Chemical Engineer's Handbook, 4th ed., McGraw-Hill Book Co., 1963.

Wicks and Block: C. E. Wicks and F. E. Block, Thermodynamic Properties of 65 Elements, Bulletin 605, Bureau of Mines, 1963.

Chapter 1

1.1 We connect two storage cells parallel. Name some extensive and some intensive properties.

1.2 We connect the two cells in series. Which properties are extensive and which are intensive?

1.3 We tie two elastic rods side by side. What can you say about strain and stress of the bundle?

1.4 We connect the rods end to end (e.g., by means of a sleeve). Which properties are extensive, which are intensive?

Chapter 3

3.1 Equation of State. Compute and plot the pressure P as a function of the molal volume $\bar{V}$ of ethane at 323.15K and 523.15K as predicted by Eq.(3.26). Compare the results with the following data (J. A. Beattie, C. Hadlock and N. Poffenberger, J. Chem. Phys., 3 (1935) 93):

$\bar{V}$(lit.mole^{-1})	0.20	0.25	0.3333	0.5	1.0	2.0
T(K)			P(atm)			
323.15	60.56	55.97	49.02	38.41	22.58	12.24
523.15	192.77	153.59	116.30	79.16	40.87	20.89

The critical data are given in Table 3.1.

3.2 Second Virial Coefficient. Compute Z from the data in the preceding example, plot Z against P and estimate the pressure at which the observations deviate from Eq.(3.23) more than 2%. Estimate a similar limit for the perfect gas equation.

3.3 Fugacity Coefficient. The curve of the fugacity coefficient as a function of pressure at the critical temperature has an inflection point. Its location can be derived from Eq.(3.62) in a Z-P-diagram. Can you give a simple geometric construction for finding the pressure at the inflection point of the fugacity coefficient?

Chapter 5

5.1 Hydrogen Peroxide Production. Can hydrogen peroxide be produced by either (or both) of the reactions

$$H_2O(\ell) + 0.5\ O_2(g) = H_2O_2(\ell) \qquad (A)$$

$$CH_3{\cdot}CH(OH){\cdot}CH_3(\ell) + O_2(g) = H_2O_2(\ell) + CH_3{\cdot}CO{\cdot}CH_3(\ell)\ ? \qquad (B)$$

For a first orientation we compute the change in free energy, assuming all substances to be in their standard states at 25°C. Actually there exists a mechanism coupling the two reactions. The combined reaction (A) + (B) is technically important.

data: LRPB

5.2 Fuel Mixture. We saturate a stream of air at the temperature t(°C) with 2,3,3,4—tetramethylpentane (nonane) under a total pressure of 1 atm. The vapor pressure p(mm mercury) of this substances is given by Antoine's equation

$$\log p = 6.860 - 1417/(214.7 + t).$$

At low temperatures the saturated stream will be too "lean" and at high temperatures too "rich" for the development of maximum power. What is the optimum temperature? A perfect gaseous mixture is assumed.

5.3 Vapor Pressure. From the information of the Gibbs free energy of formation from the elements

$CCl_4(\ell)$, $\Delta G° = -15\,600$ cal.mole^{-1}

$CCl_4(g)$, $\Delta G° = -14\,490$ cal.mole^{-1}

the vapor pressure (298 K) can be derived (Brewer).

5.4 Silver Chloride. The Gibbs free energy of formation of solid silver chloride from the elements in the standard states is given by $AgCl$, $\Delta G° = -26\,244$ cal.mole^{-1} at 298 K. Calculate the pressure of Cl_2 in equilibrium with AgCl and Ag (Brewer).

5.5 Calcium Carbonate exists in nature in two crystalline modifications, calcite and aragonite. We find for the Gibbs free energy of formation

$$\Delta G°(\text{calcite}) = -288\,460 + 62.60T + 0.896P \;(\text{cal.mole}^{-1})$$

$$\Delta G°(\text{aragonite}) = -288\,510 + 63.54T + 0.827P$$

between 298 and 450 K and for pressures up to 5000 atm. Represent the pressure at which the two modifications are in equilibrium as a function of the temperature. What is the equilibrium pressure at 300 K and 350 K?(Brewer).

Chapter 7

7.1 Azeotropic and Critical Points. Compare the properties of an azeotropic point and a critical (liquid-vapor) point of a binary mixture (simple diagrams with proper legends are in order).

7.2 Activity Coefficients from Azeotropic Data. In the system Methanol(1)-Ethyl acetate(2) an azeotropic point has been observed at 39.76°C at 446 mm mercury with $x_1 = y_1 = 0.397$. The vapor pressures are $p_1^o = 259.8$ mm, $p_2^o = 401.3$ mm. Assume that the relation

$$Q = Bx_1x_2$$

furnishes a satisfactory representation of the activity coefficients. Calculate B from the azeotropic data and compute γ_1, γ_2, P and y_2 for a few values of x_1 (0,0.25,0.5,0.75,1). Represent the calculated values of P as a function of x_1 and as a function of the calculated values of y_1 by two curves in a single diagram. Indicate the observed values:

P	437.4	446.4	384.7	301.4
x_1	0.146	0.336	0.789	0.963
y_1	0.208	0.341	0.579	0.853

7.3 Activity Coefficients. For the vapor-liquid equilibrium of mixtures of ethanol(1) and cyclohexane(2) at 5°C we find (G. Scatchard and F. G. Satkiewicz, J. Amer. Chem. Soc., 86 (1964) 130):

P(mm Hg)	16.798	36.772	48.324	48.915	48.482	36.457
x_2	0	0.0918	0.3929	0.6142	0.9041	1
y_2	0	0.5821	0.7048	0.7178	0.7299	1

We disregard the imperfection of the vapor and the Poynting correction for the liquid.

Calculate $\log \gamma_1$; $\log \gamma_2$; Q. Plot $\log(\gamma_1/\gamma_2)$ against x_1 and discuss the curve. Plot Q and indicate the calculated slopes in the endpoints and the experimental points.

7.4 Uranium. It has been suggested that fission products can be extracted from molten uranium(2) by means of molten silver(1). For the activity coefficients of the liquid mixture at 1400 K we find

$$\ell n\ \gamma_1 = 1.23x_2^2; \quad \ell n\ \gamma_2 = 1.52x_1^2 .$$

(Standards states are the pure metals.)

Calculate the solubilities of the two metals in one another (Brewer).

7.5 Iodine. In the following equations (from NBS Technical Note 270-3; 1958), valid between 270 and 350 K and below 1 atm, the molality of iodine is denoted by m (moles I_2 per kg H_2O) and by m' (moles I_2 per kg CCl_4), and its vapor pressure by p. The change ΔG in Gibbs free energy is given in cal.mole^{-1}.

$$I_2(s) = I_2(aq),\ \Delta G = 5400 - 5.04T + RT\ \ell n\ m$$
$$I_2(s) = I_2(\text{in } CCl_4),\ \Delta G = 6000 - 11.24T + RT\ \ell n\ m'$$
$$I_2(s) = I_2(g),\ \Delta G = 14\,920 - 34.52T + RT\ \ell n\ p.$$

(a) A solution containing 0.01 moles I_2 per kg CCl_4 is in equilibrium with an aqueous solution. What is the value of m at 300 K; at 350 K?

(b) What is the vapor pressure of I_2 in equilibrium with such a solution at 300 K; at 350 K?

(c) What is the value of m of an aqueous solution saturated by $I_2(s)$ at 300 K; at 350 K?(Brewer).

Chapter 8

8.1 Cuprous Ion. Calculate the equilibrium constant for the reaction
$Cu^{++} + Cu = 2Cu^+$

at 298 K and estimate the equilibrium concentration of Cu^+ in a solution containing 0.01 mole $CuCl_2$ per liter in contact with copper. Assume that the activity coefficients (a) are unity (first approximation),(b) can be represented in accord with Eq.(8.38) by

$$\log \gamma_c = -0.505\, z_c^2 \sqrt{\mu}$$

(second approximation).

data: Table 8.4.

8.2 A Fuel Cell. We suppose that a cell can be constructed in which only the reaction

$$CH_3OH(\ell) + 1.5\, O_2(g) = CO_2(g) + 2H_2O(g)$$

proceeds. What is the voltage of this cell? For a first orientation we assume substances in their standard states at 25°C.

data: LRPB

8.3 Removal of Hydrogen Sulfide. The suggestion has been made to remove hydrogen sulfide from effluent gases by oxidation to sulfur and water in a galvanic cell which would furnish electrical energy. In order to examine this suggestion, we make the following assumptions:

(a) An electrode can be found that acts practically reversibly for the conversion of hydrogen sulfide to sulfur in neutral aqueous solution. (This assumption is very questionable; satisfactory oxygen electrodes are known.)

(b) Hydrogen sulfide is removed down to a partial pressure of 10^{-4} atm.

(c) The oxygen electrode is fed with air.

(d) The cell operates at 25°C; all gases are saturated with water vapor at this temperature (vapor pressure 0.0312 atm.).

Can electric energy be obtained under these conditions. What is the reversible voltage of the cell? How much heat is evolved or absorbed in reversible operation for the conversion of 1 mole of hydrogen sulfide?

It is not likely that an electrode such as described in (a) can be found.

data: Perry's Handbook, Section 3.

8.4 Chlorine-Alkali Electrolysis. In the mercury cell for the production of sodium hydroxide and chlorine an amalgam containing about 0.2 wt.% sodium is obtained. The amalgam formation (reaction A) competes with the evolution of hydrogen (reaction B). It is technically important that practically the whole current is used for the amalgam formation.

That this objective can be attained is surprising in view of the strongly electropositive character of sodium.

(a) Write down the two complete reactions for the transport of one Faraday equivalent.

(b) Calculate the voltages of the two cells in which these reactions occur reversibly. The difference between the two voltages is the effective overpotential of hydrogen on mercury at the given conditions.

Conditions:

(1) Temperature 25°C.

(2) Electrolyte: 3 moles NaCl/1000 g H_2O (the actual concentration in technical cells is higher).

(3) Electrolyte pH = 7.

(4) 0.206 wt.% Na in the amalgam.
Na, Na(0.206 wt.% in Hg); E = 0.8456v (measurements by G. N. Lewis and C. A. Kraus, J. Amer. Chem. Soc., 32 (1910) 1459).

(5) Assume $\gamma(H^+$ in 3M NaCl) = 0.72
$\gamma(Cl^-$ in 3M NaCl) = 1.31
γ(NaCl in 3M solution) = 1.31
$Cl^- = 0.5\ Cl_2 + e^-$; $E^\circ = -1.359$
See Table 8.4.

8.5 Copper Refining. In a copper refining cell the electrolyte contains

Cu^{++}	42 grams/liter
Zn^{++}	2 grams/liter
Fe^{++}	5 grams/liter.

We assume that the activity coefficients of all divalent ions in the electrolyte are equal (sufficiently good assumption).
We assume that Zn and Fe form a perfect solution in Cu (very poor assumption).
What is the order of magnitude of the mole fractions of Zn and Fe to be expected in the refined copper?
data: Table 8.4.

8.6 Cell Capacity. We call "ideal energy capacity" the output in kW.h of a reversibly operating cell divided by the weight in kg of the reacting substances; thus we disregard the weight of leads and vessel, any necessary excess of reacting substances such as water, and any side reactions. For a comparison of the lead storage cell with the silver-zinc cell

$$0.5\ Ag_2O_2(s) + Zn + H_2O' = Ag + Zn(OH)_2(s)$$

we assume the reversible voltages to be 2.0 v (Pb) and 1.65 v (AgZn). What are the ideal energy capacities? We calculate the "ideal charge capacities" in a similar manner.

8.7 Zinc Hydroxide. We compute the standard free energy of $Zn(OH)_2(s)$ from the information given in the preceding problem, data given by Wicks and Block, and the value 0.40 assumed for the activity of water in 40% KOH solution, which serves as an electrolyte in the silver-zinc cell.

8.8 Dissociation of Acetic Acid. If a solution of an electrolyte contains an additional amount of one of the ions, we avoid the use of the degree of dissociation, which may lead to misunderstanding. Denoting the concentration of hydrogen ion by c_1, acetate ion by c_2, and undissociated acid by c_3, we write for the dissociation constant at 25°C (Section 8.62)

$$K = c_1c_2/c_3 = 1.753\times10^{-5}.$$

We prepare four solutions containing (per liter) each 0.01 mole acetic acid and in turn (a) 0.01 mole hydrochloric acid, (b) 0.001 mole sodium acetate, (c) 0.0095 mole sodium hydroxide, (d) 0.005 mole sodium chloride. What are the values of c_1, c_2, c_3 in each case? Deviations from the perfect solution may be disregarded (Brewer).

Chapter 9

9.1 Ethanol Production. Ethanol can be produced from ethylene and water at 300°C in the presence of an acidic catalyst.

(a) Compute the equilibrium constant for this process.

(b) For an approximate information of the efficiency of the process, we assume that the gaseous mixture is perfect, that the feed consists of equal molal amounts of ethylene and water, and that the pressure is maintained at 100 atm. What is the composition of the equilibrium mixture (in mole fractions)?
Would you suggest higher or lower pressure?

data: LRPB

9.2 Combustion Engine. We burn 1 mole octane to carbon dioxide and water with 5% excess over the stoichiometric amount of air. At first we calculate the approximate composition of the exhaust gas under the (rough) assumption that

(a) the combustion is complete.

Actually the exhaust contains minor amounts of pollutants, especially carbon monoxide, nitric oxide, and nitrogen dioxide. For an estimate of the mole fractions of these minor constituents, we assume that

(b) we may neglect the changes in the amounts of major constituents, caused by the formation of the pollutants (i.e., the mole fractions calculated under assumption (a) for the major constituents can be retained without change),

(c) equilibrium between major and minor constituents of the exhaust gas is reached at 1000 K and 50 atm,

(d) gas imperfection may be disregarded.

What is the approximate composition of the exhaust gas? Which pollutants are increased by a rich mixture (insufficient air supply) or by a lean mixture (excess air)?

data: Wicks and Block

9.3 Aluminum Production. The main reaction in the electrolytic reduction of alumina at 1250 K is

$$Al_2O_3(s) + 1.5\ C(graphite) \ = \ 2Al(\ell) + 1.5\ CO_2 \ .$$

Calculate the minimum voltage required and the minimum electric energy (in kW.h per kg Al).

data: LRPB

9.4 Side Reaction in Aluminum Production. The appearance of carbon monoxide

$$Al_2O_3(s) + 3\ C(graphite) \ = \ 2Al(\ell) + 3CO$$

is undesirable. Calculate minimum voltage and energy (in kW.h per kg Al).

data: LRPB

9.5 Total Energy in Aluminum Production. Calculate the total energy consumed by each of the two preceding reactions and express it in kW.h per kg Al. The practical consumption is 11-16 kW.h/kg Al.

data: K. K. Kelley, Bureau of Mines, Bulletin 594 (1960).

9.6 Argon in Ammonia Synthesis. Atmospheric nitrogen contains about 1% Argon, which is enriched on recycling. Calculate the conversion of a feed containing n_1 = 1 mole N_2, n_2 = 1.5 moles H_2, n_4 moles Ar at 450°C and pressures up to 1000 atm. The influence of the fugacity coefficients can be approximated by Tables 9.1 and 9.2. Plot the conversion for 100 and 1000 atm and n_4 between 0 and 0.2.

9.7 Explosion Danger. Which conclusions can we derive from thermodynamic data concerning the danger of an explosion of a mixture consisting of

(a) benzene and hydrogen peroxide,

(b) ethanol and hydrogen peroxide,

(c) nitrogen and ammonia,

(d) nitrogen and water?

data: LRPB

9.8 Nitric Oxide. The reaction

$$0.5\ N_2 + 0.5\ O_2 = NO$$

has played an important role in various problems. Compute the equilibrium fraction of NO at 1000 K, 2000 K, and (by a suitable, bold extrapolation) at 10 000 K. Discuss briefly: (a) Attempts at direct fixation of nitrogen at 2000 K. (b) Formation of nitric oxide in combustion engines and smog. (c) Danger of burning up the atmosphere in a nuclear explosion.

data: LRPB

9.9 Ozone. (a) Calculate the partial pressure of ozone in equilibrium with the oxygen of the atmosphere at 25°C. (b) In commercial ozonizers about 50g ozone at a partial pressure of 0.1 atm in oxygen at atmospheric pressure is obtained per kW.h. What is the efficiency (minimum/actual electric energy required).

data: Perry's Handbook, Section 3.

9.10 Ammonia Oxidation. We mix ammonia with air and obtain equilibrium with nitric oxide over a catalyst at 1250 K and 1 atm. An excess of 20% over the stoichiometric amount of oxygen is used. What is the residual fraction of ammonia in the effluent?

data: Wicks and Block.

9.11 Hydrogen Burner. Estimate the highest temperature attainable by burning hydrogen in air. For a rough estimate we may assume that the molal heat capacity of water vapor is the same as that of nitrogen.

data: LRPB

9.12 Carbon Tetrachloride Purification. Is the refluxing of carbon tetrachloride with mercury a safe procedure? It is sufficient to use the data for 25°C without converting to the boiling temperature.

data: LRPB

9.13 Acetylene. Acetylene can be produced by rapid heating of methane and subsequent rapid cooling. Decomposition of the gas to graphite and hydrogen is a source of losses. What is the composition of the product in each of the reactions

$$2CH_4 = C_2H_2 + 3H_2$$
$$C_2H_2 = H_2 + C(\text{graphite})$$
$$CH_4 = 2H_2 + C(\text{graphite})$$

if equilibrium is reached at 1500 K and 1 atm? Show that the manufacture of acetylene would be impossible if the second and third reactions take place to an appreciable extent.

data: LRPB

9.14 Sulfuric Acid. Prepare a table similar to Table 9.4 for the production of sulfuric acid from pyrite.

9.15 Phosgene. Is there any danger in the use of carbon tetrachloride in fire extinguishers? For a first orientation we consider the following reactions at 1000 K:

$$CCl_4 + CO_2 = 2COCl_2 \qquad (A)$$
$$CCl_4 + C(\text{graphite}) + O_2 = 2COCl_2. \qquad (B)$$

We assume that the reacting gases are present at the partial pressures of 0.1 atm. What would be the partial pressures of phosgene if equilibrium were reached?

data: LRPB

9.16 Sucrose can hydrolyze according to

$$C_{12}H_{22}O_{11}(\text{sucrose}) + H_2O = C_6H_{12}O_6(\text{glucose})$$
$$+ C_6H_{12}O_6(\text{fructose}); \quad \Delta G^\circ = 7000 \text{ cal}$$

Calculate the equilibrium constant and the hydrolyzed equilibrium fractions at 25°C for initial concentrations of 10^{-3}; 10^{-5}; 10^{-6} mole sucrose per liter (Brewer).

Symbols

Molal quantities are denoted by capitals carrying a bar. A page number in parentheses means that the symbol is used only in a small part of this book.

a	activity parameter in equations of state (attraction coefficient) Stefan-Boltzmann parameter (238)
A	free energy of Helmholtz interaction coefficient integration constant (164) area (227)
A,B,C,D,	coefficients of vapor pressure equations (94) abbreviation terms (203)
A',B'	coefficients of van Laar's equation (112)
A",B",C"	coefficients of Wilson's equation (118)
A_s	surface area (214)
b	parameter in equations of state (limiting volume)
B	coefficient in adsorption equations (220)
B	magnetic induction (234)
B,C,D	coefficients in a power series (109)
c	number of independent components (87) concentration (molarity) conversion (195)
C_v,C_p	heat capacity
D	dielectric constant (162,228) Debye parameter (238)
E	energy
e	electric field strength (228)
E	electromotive force
f	generalized force degrees of freedom (87) centrifugal force (224)
F	final state (13) dampening factor (204)
F	Faraday equivalent
g	statistical weight (68,244) exponent in adsorption equations (220) gravitational constant (224)
G	Gibbs free energy an extensive quantity (16)
h	Planck's constant (74) Henry's coefficient (130) an interaction factor (166)
H	heat content, enthalpy magnetic field strength (232)
I	initial state (13)
k	Boltzmann's constant coefficient for the molal volume (169)
K	equilibrium constant
ℓ	distance between plates
L	relative heat content (145) length (214)
m	molality (144)
M	molal weight, mass
M	magnetization (231)
n	number of moles
N	number of molecules

N_L	Avogadro's constant (Loschmidt's number per mole)
p	number of phases (87) vapor pressure, partial pressure linear momentum (237)
P	pressure
q	heat (24) a freezing point function (148) a quotient (197) amount of adsorbate (219) cross section (224)
$\mathbf{q}$	dielectric displacement (228)
Q	an arbitrary variable (8) partition function (69) a dimensionless excess quantity (106)
$\mathbf{Q}$	electric charge
r	radius (218) distance from center (162,224) an abbreviation term (199)
R	gas constant
s	specific weight (160)
S	entropy
t	temperature (°C)
T	temperature
$u = hc\bar{\nu}/kT$ (75)	radiation energy density (237)
v	velocity
$\bar{v}_i$	partial specific volume
V	volume
w	work weight of solvent (149) valence factor (166) weight in error calculations (253)
W	probability (German:Wahrschein-lichkeit) free energy of interaction (165)
x	generalized coordinate mole fractions
y	mole fractions (especially for gaseous mixtures)
z	ionic charge
Z	compressibility factor (46)
α	separation factor (121) degree of dissociation (181)
β	second virial coefficient compressibility "true" activity coefficient(181)
γ	activity coefficient
Γ	surface concentration (215)
δ	solubility parameter (113)
ε	molecular energy (68) electron charge (161) permittivity (228) a small error (253)
η	measure of irreversibility (35) ratio O_2/S (198)
θ	temperature functions (30,148) freezing point depression
κ	interaction coefficient (163)
λ	a temporary parameter(165)
Λ	molal conductivity
μ	Joule-Thomson coefficient (53) chemical potential interaction coefficient (115) ionic strength (160) permeability (234)
ν	frequency split factor (158) reaction coefficient (172)
$\bar{\nu}$	wave number
Π	osmotic pressure (153)
ρ	charge density (162) an abbreviation term (195) density (225)
σ	surface tension
τ	moment of torsion (231)
ϕ	fugacity coefficient apparent molal volume (169) angle of deflection (231)
Φ	interaction potential (164)
ψ	electrostatic potential (162)
ω	acentric factor (52) angular velocity (224) an angle (231)

Author Index

Subject Index